A Simple Approach To Group Theory IN Chemistry

S Swarnalakshmi
Reader, Guru Nanak College, Chennai

T Saroja
Selection grade lecturer, Guru Nanak College, Chennai

R M Ezhilarasi
Senior scale lecturer, Guru Nanak College, Chennai

Universities Press

A SIMPLE APPROACH TO GROUP THEORY IN CHEMISTRY

UNIVERSITIES PRESS (INDIA) PRIVATE LIMITED

Registered Office
3-6-747/1/A and 3-6-754/A Himayatnagar, Hyderabad 500 029, Telangana, India
email: info@universitiespress.com

Distributed by
Orient Blackswan Private Limited

Registered Office
3-6-752 Himayatnagar, Hyderabad 500 029, Telangana, India

Other Offices
Bengaluru, Chennai, Guwahati, Hyderabad, Kolkata,
Mumbai, New Delhi, Noida, Patna, Visakhapatnam

First published in India by
Universities Press (India) Private Limited 2008
Reprinted 2008, 2009, 2012, 2015, 2019, 2020, 2023

501912

ISBN 978-81-7371-623-2

Typeset at
Trinity Designers & Typesetters, Chennai 600 041

Printed at
B.B. Press, Noida 201 301

Published by
Universities Press (India) Private Limited
3-6-747/1/A and 3-6-754/A Himayatnagar, Hyderabad 500 029, Telangana, India.

Foreword

Group theory – a topic that most students of chemistry find difficult at the postgraduate level – forms an integral part of understanding chemistry, with emphasis on the structure of molecules. One cannot claim mastery of chemistry without having a thorough understanding of symmetry and group theory, and its applications to chemistry.

The authors have more than two decades of teaching experience in handling undergraduate and postgraduate students of the Madras University. They have chosen to present the various aspects of group theory (relevant to chemistry) in a simple and easily understandable form for students who lack expertise in mathematics. An important aspect of this book is the clarity of presentation.

The first four chapters introduce the basic concepts in group theory in a simple and logical manner with a large number of examples. The student will have no difficulty in visualising the various representations of symmetry operations. The applications of group theory to spectroscopy are dealt with in chapters 6 to 9. A thorough understanding of these chapters will form a good introduction to specialised areas of spectroscopy.

This book is intended as a reference book for undergraduates of chemistry and as a textbook for postgraduates in Indian Universities. It will also serve as useful study material for students who wish to appear for the UGC-NET, IAS and other All India based competitive examinations.

I congratulate the authors for having chosen to device a text of this kind where the fundamentals of group theory have been neatly presented, laying a strong foundation for understanding the applications of group theory in chemistry.

S JAYARAMAN
Professor and Head (Retd),
Department of Chemistry,
Guru Nanak College,
Chennai 600 042

Acknowledgements

We thank the management, Principal and faculty members of the chemistry department of Guru Nanak College, Chennai, for their constant encouragement and support during the preparation of the manuscript. We are grateful to Dr S Jayaraman, former Principal, Guru Nanak College, and Prof S Srivathsan, former Head, Department of Chemistry, Guru Nanak College, for motivating and inspiring us to write the book. Thanks are due to the editor Dr Gita S Dattatri for her cooperation and fullest assistance in this endeavour. Finally, we thank our family members for their cooperation and support during the preparation of the manuscript. Suggestions from teachers and students for the improvement of the book will be thankfully received.

AUTHORS

Contents

1

Symmetry Elements and Symmetry Operations

1.1 Importance of Symmetry

Symmetry is a familiar concept in our world. Symmetry means that a certain portion of an object looks exactly like another portion of the same object. In nature, we find countless examples of symmetry. Many animals have bilateral symmetry which means, the two halves are mirror images of each other. This is equivalent to them possessing a plane of symmetry. Symmetry is exhibited in the plant kingdom in the structure of leaves and flowers. It is also observed in the formation of crystals which may take the shape of a cube, an octahedron or an icosahedron. Symmetry is an integral part of arts like music, painting and designing because of the need for aesthetic regularity, pleasing proportions, periodicity and harmonious arrangement. The involvement of symmetry in chemistry has been known and studied for a long time. In 540 BC, the society in which Pythagoras lived held the view that the earth had been produced from a regular hexagon or cube, fire from a regular tetrahedron, air from a regular octahedron, water from a regular icosahedron and the heavenly sphere from a regular dodecahedron!

Symmetry plays an important role in molecular spectroscopy, in quantum mechanics as well as in the determination of structure of molecules and understanding of stereochemistry. A thorough knowledge of the structure, shape and symmetry of molecules and an overall understanding of the symmetry and mechanisms of reactions leads to efficient designing of asymmetric syntheses which are vital in chemical, pharmaceutical and polymer industries.

The systematic discussion of symmetry is called *group theory*. This theory gives a simple and direct method for arriving at useful conclusions with minimum calculations about the geometry and electronic structure of molecules. The frontier molecular orbital theory and the correlation method which are useful in predicting the 'allowedness' of the synthetically challenging pericyclic reactions are mainly based on group theory. The ligand field theory is also supported by group theory.

To understand the various symmetry elements present in a molecule, it is necessary to have some basic knowledge about the geometry of various molecules. This can be obtained from the *Valence Shell Electron Pair Repulsion Theory (VSEPR).*

1.2 VSEPR Theory to Predict Geometry of Molecules

This theory was developed by Gillespie to predict and explain molecular geometry and bond angles. It can be summarised as follows.

- The central atom is first identified. The shape of the molecule is determined by considering the repulsions between all the electron pairs in the valence shell of the central atom.
- A lone pair is confined to a single nucleus whereas a bond pair is shared by two nuclei. Hence, the non-bonding electron pairs occupy more space around the central atom than the bonded pair of electrons.
- Therefore, the repulsion between lone pairs of electrons is greater than that between two bonded pairs of electrons.
- The magnitude of repulsion between bonding pairs of electrons depends on the electronegativity difference between the central atom and the other atoms.
- Double bonds cause more repulsion than single bonds and triple bonds in turn cause more repulsion than double bonds. Molecules assume a geometry with minimum repulsion.

Depending on the number and nature of electron pairs in the valence shell, shapes of molecules can be predicted by applying the VSEPR theory. The number and nature of electron pairs present in molecules also help in predicting the bond angles of these molecules. Table 1.1 indicates the correlation between the number of electron pairs and the shape of molecules.

1.3 Symmetry Elements

Symmetry elements are defined as imaginary geometrical entities such as points, lines and planes that are present in a molecule, about which when symmetry operations are performed, the molecule presents an indistinguishable configuration. The nature of symmetry elements present in any molecule depends on its geometry.

Table 1.1 Effect of bonding and lone pairs of electrons on shape and bond angles of molecules

Molecule	No. of hybrid orbitals on central atom	Shape	No. of bond pairs	No. of lone pairs	Bond angle
$BeCl_2$	2	Linear	2	0	180°
BF_3	3	Triangular planar	3	0	120°
CH_4	4	Tetrahedral	4	0	109°48′
NH_3	4	Tetrahedral	3	1	107°48′
NF_3	4	Tetrahedral	3	1	102°30′
H_2O	4	Tetrahedral	2	2	104°27′
F_2O	4	Tetrahedral	2	2	102°
PCl_5	5	Trigonal bipyramidal	5	0	120° , 90°
SF_4	5	Trigonal bipyramidal	4	1	101°36′, 86°33′
ClF_3	5	Trigonal bipyramidal	3	2	87°40′
I_3^-	5	Trigonal bipyramidal	2	3	180°
SF_6	6	Octahedral	6	0	90°
BrF_5	6	Octahedral	5	1	84°30′
XeF_4	6	Octahedral	4	2	90°

1.3.1 What are indistinguishable configurations?

These are spatial arrangements which cannot be distinguished from one another. The two configurations are alike in all respects (but are not exactly identical) and it is difficult to predict if any change has been carried out or not. However, they can be differentiated by introducing appropriate suffixes in the form of numbers or letters. For example, configurations I and II (Fig. 1.1) are indistinguishable but for the suffixes 1 and 2 on the B atoms.

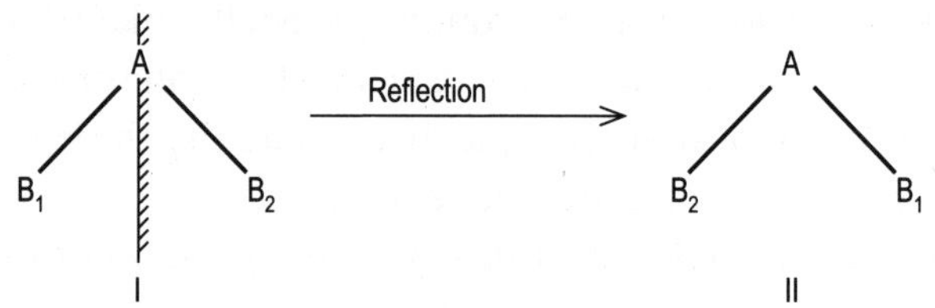

Fig 1.1

1.3.2 Symmetry operations

These are simple geometric operations such as reflection, rotation or inversion which when performed on the molecule, give rise to an indistinguishable configuration of the same molecule. Thus, the presence of symmetry elements in a molecule can be identified by performing the appropriate symmetry operations (Table 1.2).

Table 1.2 Types of symmetry elements and their corresponding symmetry operations

Symmetry element	Corresponding symmetry operation
Identity element (E) or (I)	doing nothing (360° rotation)
Plane of symmetry (σ)	reflection
Centre of symmetry (i)	inversion of all coordinates
Proper axis of symmetry (C_n)	rotation through an angle of $\left(\frac{360}{n}\right)^{\circ}$ about an axis where n is the order of the axis.
Improper axis of symmetry (S_n)	rotation through an angle of $\left(\frac{360}{n}\right)^{\circ}$ about an axis followed by reflection about a plane perpendicular to the original axis.

Identity element (E)

The identity element is present in all molecules. Identity is the operation of not doing anything. When a molecule is not operated upon by any symmetry element, the system is left unchanged and the configuration of the molecule is identical with the original configuration. Thus, the molecule is said to possess the identity element. It is denoted by the symbol E. Though it may appear that this element does not have any significance, its mathematical relevance will be clear as we proceed further.

Plane of Symmetry (σ)

When reflection is carried out about an imaginary plane in a molecule and if the molecule presents an indistinguishable configuration, the plane is called a plane of symmetry. It is represented by σ. A molecule may have one or several planes of symmetry or may not have any such plane. Let us identify the symmetry planes present in molecules possessing different geometries.

Note: All linear molecules have an infinite number of planes of symmetry.

AB_2 (Bent)

All molecules of this type possess two planes of symmetry. An example of such a molecule is H_2O.

(i) One plane of symmetry passes through atom A and bisects the line joining the two B atoms which are identical (Fig. 1.2). But for the subscripts, the atoms B_1 and B_2 in the two configurations cannot be distinguished.

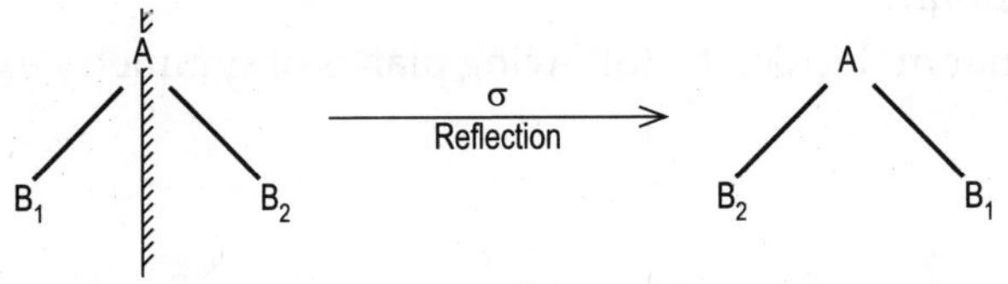

Fig. 1.2

(ii) The molecular plane: It is the plane in which the molecule lies. When reflection is done about this plane, the molecule reflects onto itself and presents an indistinguishable configuration.

Note: All planar molecules have the molecular plane as one of the planes of symmetry.

AB_3 (Planar)

A trianglular planar molecule will have the following planes of symmetry.

(i) A molecular plane (since it is planar);

(ii) A plane passing through one of the B atoms (say B_1) and A which bisects the line joining the other two B atoms (B_2 and B_3). Similarly, there are two other planes of symmetry, that is, a plane passing through A and B_2 which bisects the line joining B_1 and B_3, and a third plane passing through A and B_3 which bisects the line joining B_1 and B_2 (Fig. 1.3). An example of this type of molecule is BCl_3.

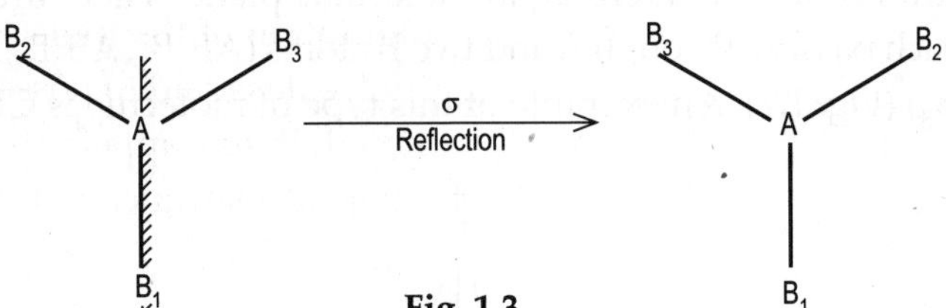

Fig. 1.3

Thus, a tetra atomic AB_3 planar molecule has four (1+ 3) planes of symmetry.

AB_3 (Pyramidal)

This type of molecule has three planes of symmetry. Each plane passes through A and one of the B atoms (B_1, B_2 or B_3) and bisects the line joining the other two B atoms (Fig. 1.4). There is no molecular plane of symmetry. An example of this type of molecule is NH_3.

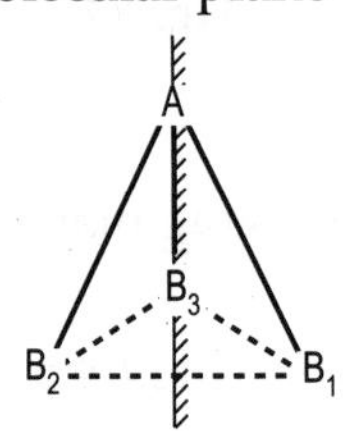

Fig. 1.4

AB_4 (Square planar)

In a square planar molecule, the following planes of symmetry exist (Fig. 1.5 and Fig. 1.6).

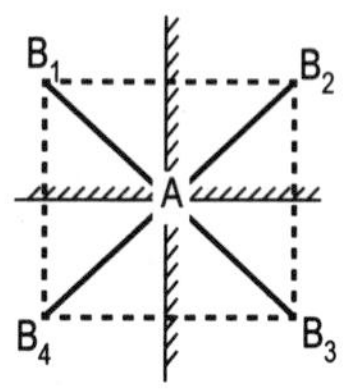

Fig. 1.5

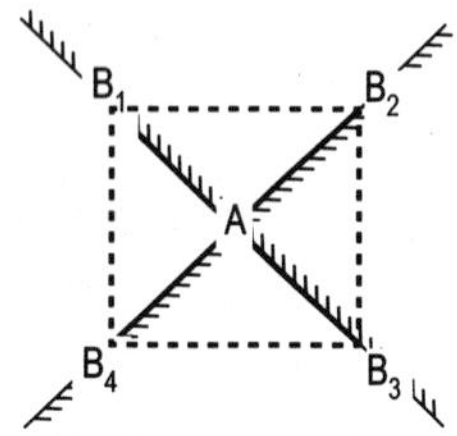

Fig. 1.6

(i) A molecular plane (since the molecule is planar);
(ii) Two planes of symmetry exist, each passing through atom A and bisecting the line joining the adjacent B atoms on either side.
(iii) Two diagonal planes each passing through A and the two diagonally opposite B atoms.

Thus, such a molecule will possess a total of five (1 + 2 + 2) planes. An example of this type of molecule is $[PtCl_4]$.

AB_4 (Tetrahedral)

In a tetrahedral molecule, there is no molecular plane. There are six planes of symmetry, each passing through A and two B atoms (AB_1B_2, AB_2B_3, AB_3B_4, AB_4B_1, AB_2B_4, AB_1B_3) (Fig. 1.7). An example of this type of molecule is CH_4.

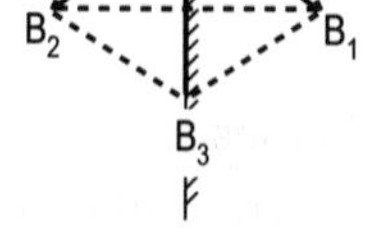

Fig. 1.7

AB_5 (Pentagonal)

In a pentagonal molecule, the following planes of symmetry exist (Fig. 1.8).

(i) Molecular plane;
(ii) Five symmetry planes each passing through A and one of the B atoms and bisecting the line joining adjacent B atoms on the opposite face.

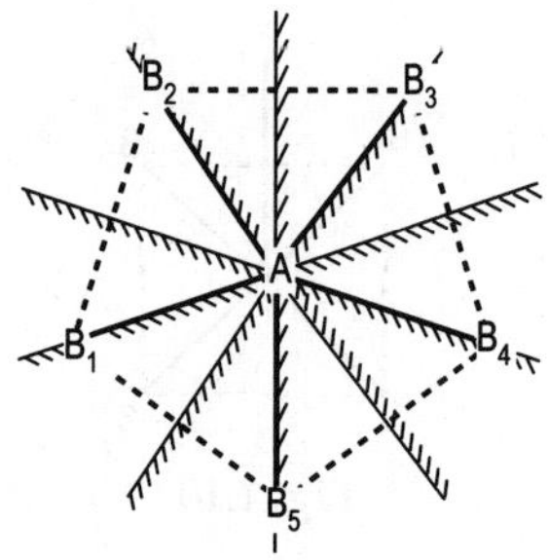

Fig. 1.8

Thus, the a molecule has a six (1 + 5) planes of symmetry. This type of molecule is hypothetical.

AB_5 (Trigonal bipyramidal)

A trigonal bipyramidal molecule has the following planes of symmetry (Fig. 1.9).

(i) A plane of symmetry passing through A and the three equatorial B atoms (B_1, B_2, B_3);

(ii) Three planes of symmetry, each passing through A, the two axial B atoms (B_4 and B_5) and one equatorial B atom.

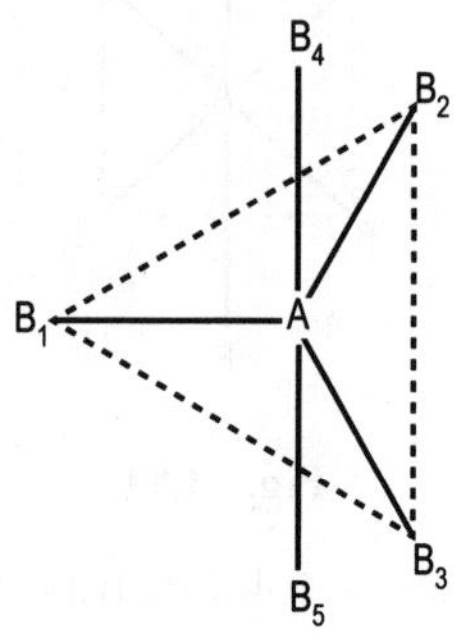

Fig. 1.9

Thus, the molecule has four (1 + 3) planes of symmetry. An example of such a molecule is PCl_5.

AB_5 (Square pyramidal)

In a square pyramidal molecule, the planes of symmetry that exist are as follows (Fig. 1.10).

(i) Two planes of symmetry, each passing through A and the axial B (B_1) which bisects the line joining adjacent B atoms on both sides;

(ii) Two diagonal planes of symmetry, each passing through A, the axial B and the two diagonally opposite equatorial B atoms.

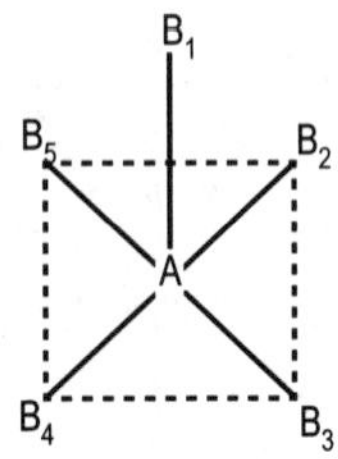

Fig. 1.10

Thus, the molecule has four (2 + 2) planes of symmetry. An example of such a molecule is IF_5.

AB_6 *(Octahedral)*

In an octahedral molecule, the following planes of symmetry exist (Fig. 1.11).

(i) Six planes passing through the opposite edges of the octahedron;

(ii) Three planes passing through A and the four equatorial B atoms and bisecting the line joining the axial B atoms.

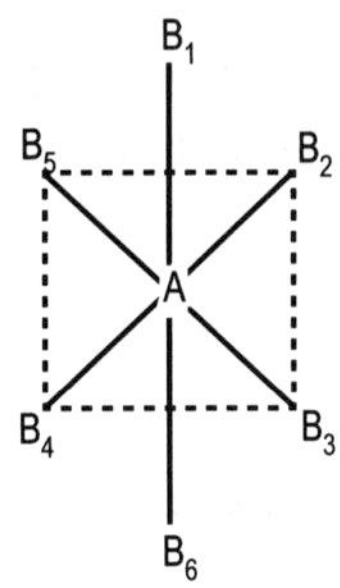

Fig. 1.11

Thus, the molecule will have a total of nine (6 + 3) planes of symmetry. An example of this type of molecule is $[Cr(CO)_6]$.

AB_6 *(Planar)*

Such a molecule does not exist. It has been used for the purposes of learning. It has the following planes of symmetry.

(i) A molecular plane (since it is planar);

(ii) Three symmetry planes, each passing through A and bisecting the line joining adjacent B atoms on both sides;

(iii) Three symmetry planes each passing through A and bisecting the line joining two opposite B atoms.

Thus, the molecule will have a total of seven (1 + 3 + 3) planes of symmetry (Fig. 1.12).

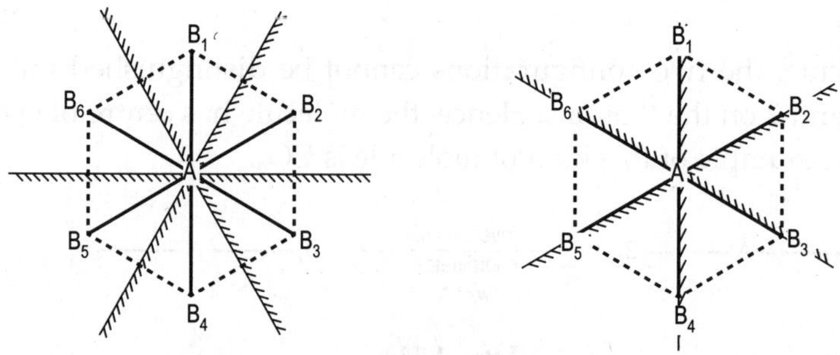

Fig. 1.12

In any molecule, the first reflection about any symmetry plane leaves the molecule in an indistinguishable configuration. The second reflection about the same plane leads to the original configuration. For example, consider the AB_2 bent molecule (Fig. 1.13). When reflection is performed along the plane passing through A and bisecting the line joining the two B atoms, the molecule attains an indistinguishable configuration (II). When the same operation is performed for the second time it gives rise to the original configuration. This is equal to the identity element, *E*.

$$\therefore \sigma^2 = E$$

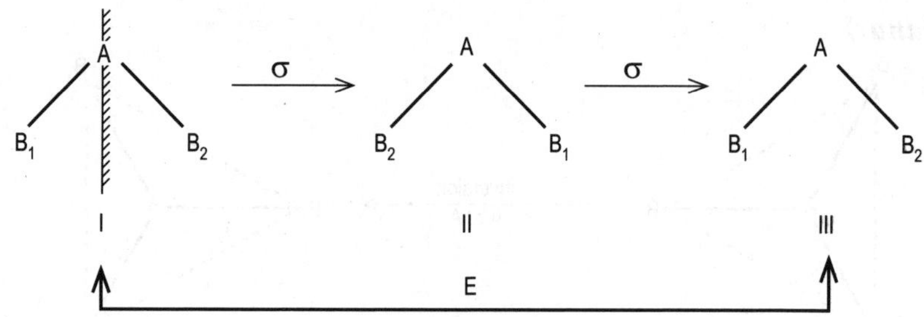

Fig. 1.13

Centre of symmetry (i)

When the coordinates (x,y,z) of every atom in a molecule are changed into (–x,–y,–z) and even after that, if the molecule presents an indistinguishable configuration, then the point of origin (0,0,0) is called the *centre of symmetry* of the molecule. That is, if an imaginary line is drawn from an atom to the centre of the molecule and extended on the other side by the same distance and meets a similar atom, the molecule is said to possess a centre of symmetry or a *centre of inversion*. For example a cube, a sphere and an octahedron possess a centre of symmetry whereas a triangle and a pyramid do not.

Let us identify the centre of symmetry in molecules of different geometries. After inversion, atoms generally shift to their exact diagonally opposite positions. So, whenever diagonally opposite atoms are the same in a molecule, the molecules will have a centre of symmetry and vice versa.

AB_2 (Linear)

In this molecule, the two configurations cannot be distinguished but for the subscripts 1 and 2 on the B atoms. Hence, the molecule has centre of symmetry (Fig. 1.14). An example of this kind of molecule is CO_2.

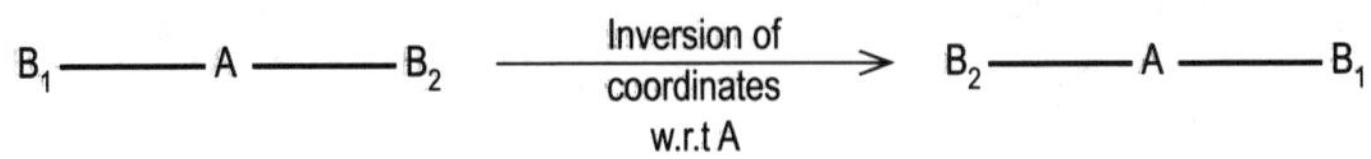

Fig. 1.14

AB_2 (Bent)

In a bent molecule of this type, when inversion is carried out, the two configurations are distinguishable and hence the molecule does not have a centre of symmetry (Fig. 1.15). An example of this kind of molecule is H_2O.

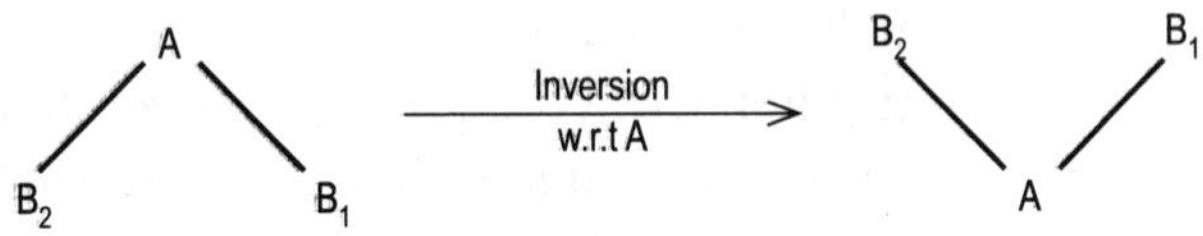

Fig. 1.15

AB_3 (Planar)

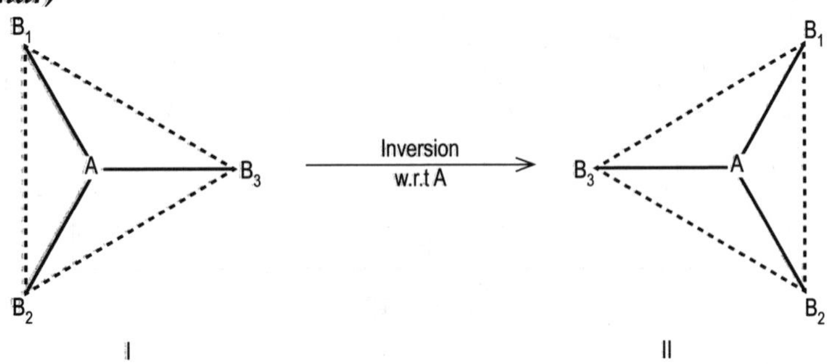

Fig. 1.16

Structures I and II are distinguishable. Hence this type of molecule does not possess a centre of symmetry. An example of this kind of molecule is BF_3.

AB_3 (Pyramidal)

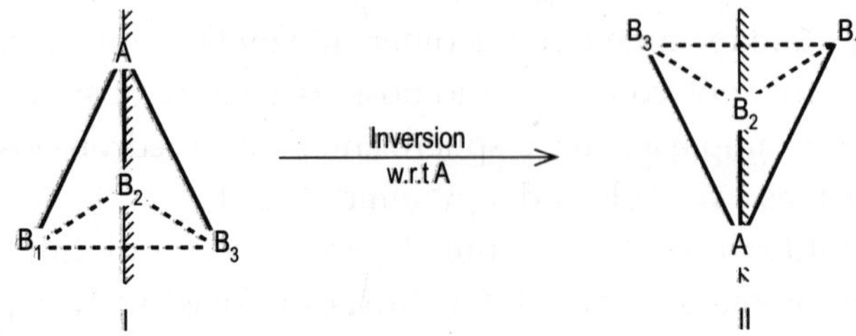

Fig. 1.17

Structures I and II are distinguishable, therefore the molecule does not possess a centre of symmetry. An example of this kind of molecule is NF_3.

AB_4 (Square planar)

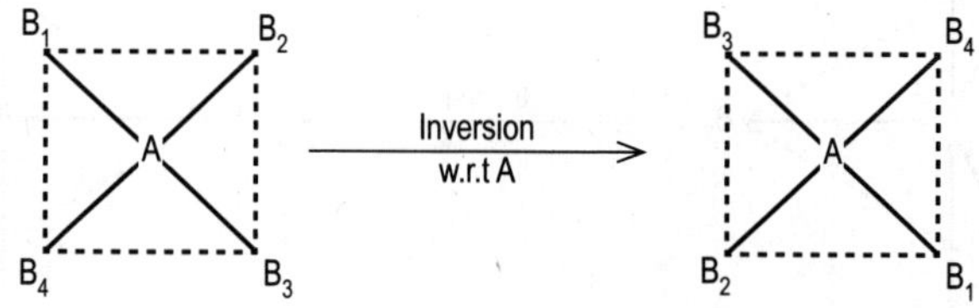

Fig. 1.18

Such a molecule presents an indistinguishable configuration when inversion is carried out (note that the diagonally opposite atoms are same). Hence, the molecule has a centre of symmetry. An example of this kind of molecule is XeF_4.

AB_4 (Tetrahedral)

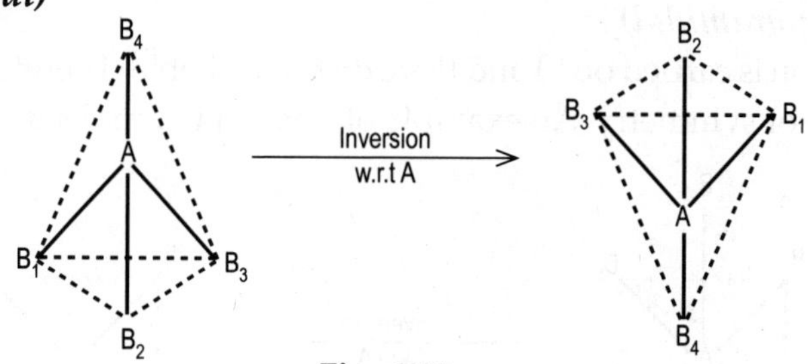

Fig. 1.19

When inversion is carried out, the structures are distinguishable. Hence, the molecule has no centre of symmetry. An example of this kind of molecule is CCl_4.

AB_5 (Pentagonal)

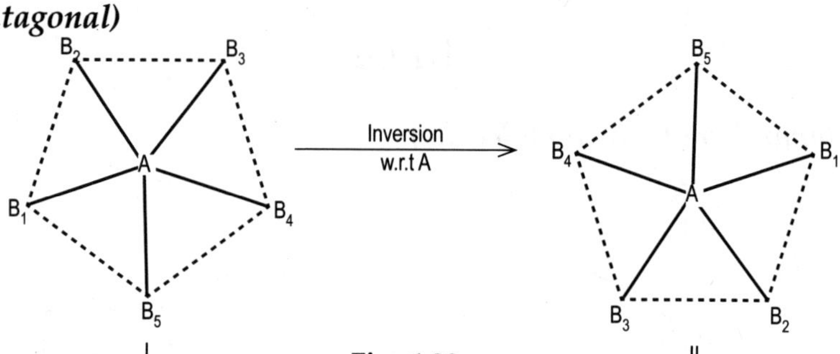

Fig. 1.20

Structures I and II are distinguishable after the inversion operation is carried out. Hence, the molecule does not possess a centre of symmetry. This kind of molecule is not known to exist.

AB_5 (Trigonal bipyramidal)

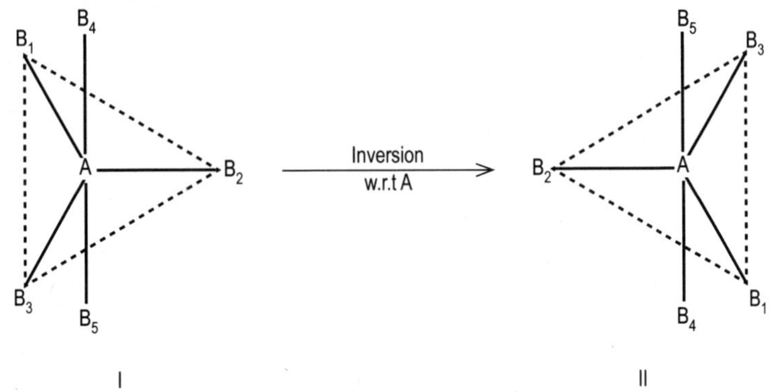

Fig. 1.21

Molecules of this type have no centre of symmetry as the configurations I and II are distinguishable. An example of this kind of molecule is $Fe(CO)_5$.

AB_5 (Square pyramidal)

When inversion is carried out, I and II are distinguishable. Therefore, the molecule has no centre of symmetry. An example of this kind of molecule is BrF_5.

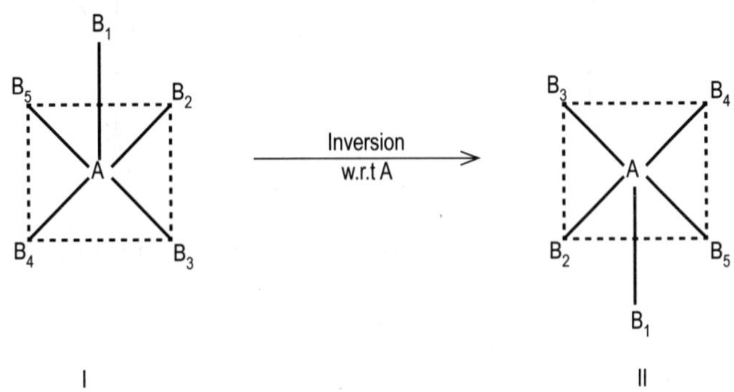

Fig. 1.22

AB_6 [Planar (hypothetical molecule)]

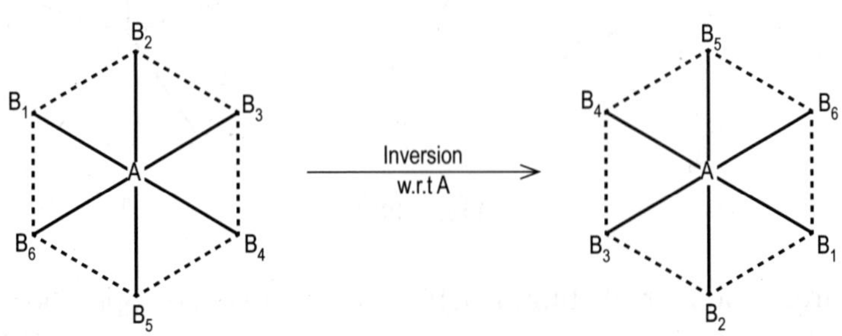

Fig. 1.23

Such a molecule does not exist in reality. However, when inversion is carried out, the molecule presents an indistinguishable configuration. Hence, the molecule has a centre of symmetry.

AB_6 (Octahedral)

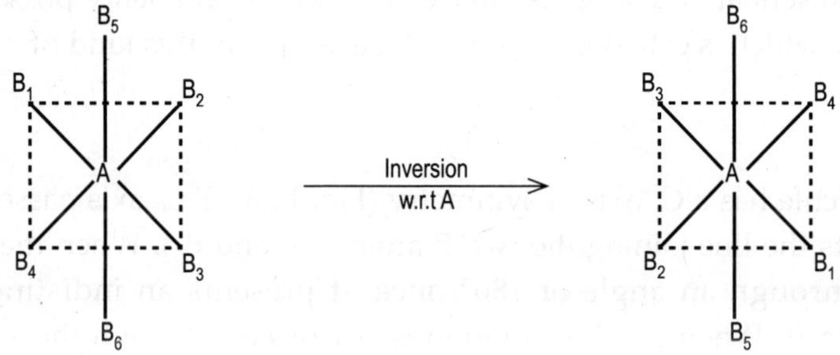

Fig. 1.24

When inversion is carried out on this molecule, the configurations I and II are indistinguishable. Hence, the molecule has a centre of symmetry. Example, SF_6. In this molecule, when inversion is performed for the first time, an indistinguishable configuration is obtained and when the same operation is performed for the second time, the molecule attains its original configuration. It can be represented as follows.

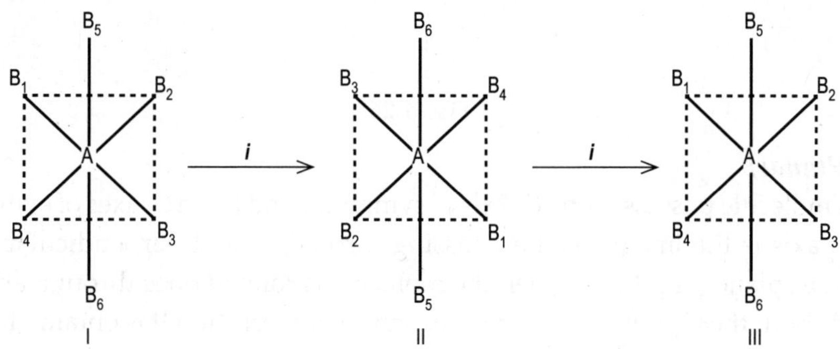

Fig. 1.25

$\therefore \qquad i^2 = E$

This relationship is true for all molecules which have a centre of symmetry.

Proper axis of symmetry (C_n)

A proper axis of symmetry is defined as an imaginary axis present in a molecule about which, when the molecule is rotated through an angle of $(360/n)°$, the molecule presents an indistinguishable configuration. Here, n represents the order of the axis and can take whole number values. When $n = 2$, the axis is referred to as the C_2 or the *diad axis;* when $n = 3$, the axis is referred to as the C_3 or the *triad axis* and so on.

The nature of C_n axis present in a molecule varies with its geometry. Let us consider different types of molecules and identify the axes of symmetry present in them.

All linear molecules, when rotated through any angle with respect to their bond axis, present an indistinguishable configuration and hence possess an axis of symmetry which is called the C_∞ axis. An example of this kind of molecule is HCl.

***AB$_2$* (Bent)**

Such a molecule has a C_2 axis of symmetry (Fig. 1.26). This axis passes through A and bisects the line joining the two B atoms (B_1 and B_2). When the molecule is rotated through an angle of 180° once, it presents an indistinguishable configuration II. When another rotation is performed through the same angle 180°, it leads to the original configuration I. That is, $C_2^2 = E$. An example of this kind of molecule is H_2S.

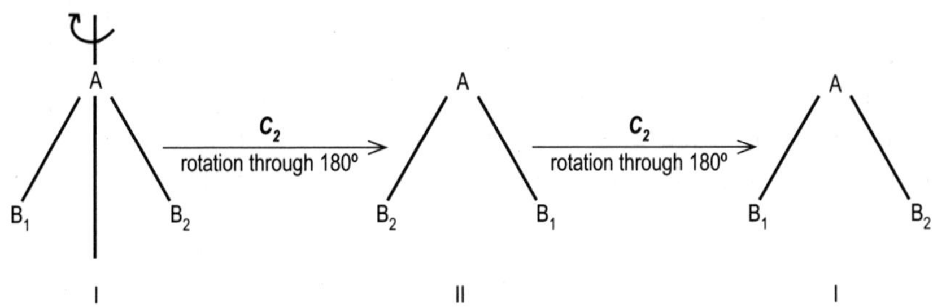

Fig. 1.26

***AB$_3$* (Planar)**

Such a molecule possesses one C_3 axis of symmetry and three C_2 axes of symmetry. The C_3 axis is the imaginary line passing through A and perpendicular to the molecular plane (Fig. 1.27). When the molecule is rotated once through an angle of 120° about the C_3 axis, an indistinguishable configuration II is obtained. When a similar rotation is performed on II through an angle of 120°, it leads to another indistinguishable configuration III. A further rotation of III through an angle of 120° leaves the molecule with the original configuration IV.

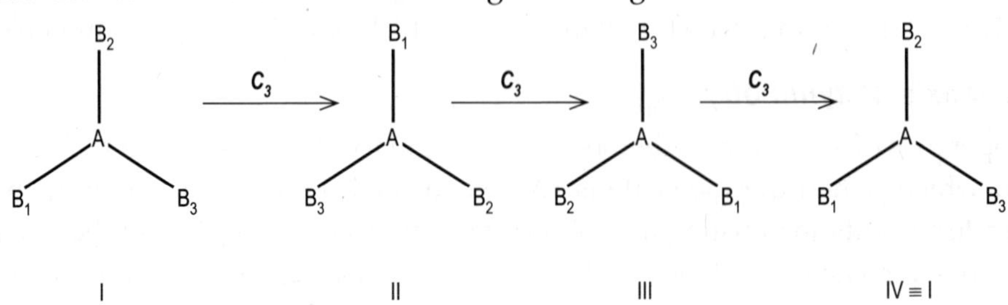

Fig. 1.27

That is, $C_3^3 = E$.

The C_2 axis is the imaginary line passing through A and one of the B atoms (say, B_1) and bisecting the line joining the other two B atoms (B_2 and B_3). There are three such axes present in the molecule (Fig. 1.28). These axes lie on the molecular plane and are perpendicular to the C_3 axis. An example of this kind of molecule is BCl_3.

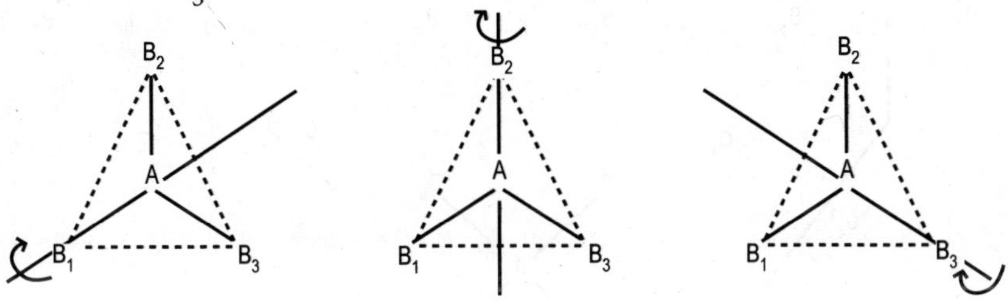

Fig. 1.28

AB_3 (Pyramidal)

Such a molecule possesses only the C_3 axis of symmetry. It passes through A and the centre of the triangular plane whose corners are occupied by the three B atoms (Fig. 1.29).

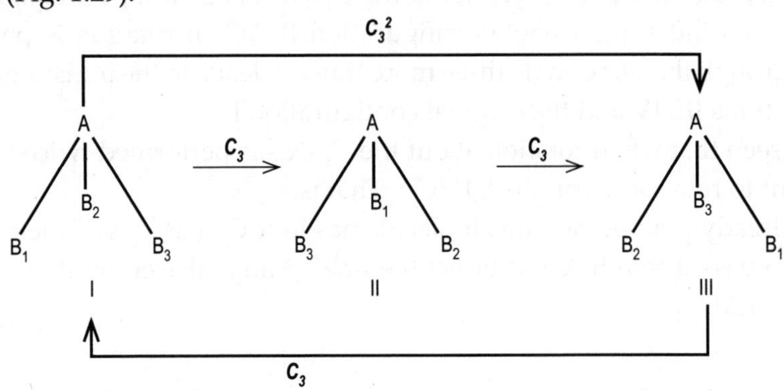

Fig. 1.29

When rotation is performed on this molecule once, twice and thrice through an angle of 120°, it leads to indistinguishable configurations II and III first, and then the original configuration is obtained again. Thus, $C_3^3 = E$. This molecule has no C_2 axis of symmetry and no molecular plane of symmetry. An example of this kind of molecule is NH_3.

Note: Any C_n axis is associated with *n* number of symmetry operations or symmetry elements. They ae C_n, C_n^2....C_n^n.

AB_4 (Square planar)

This molecule has one C_4 axis of symmetry and four C_2 axes of symmetry. The C_4 axis is the imaginary line passing through A and perpendicular to the molecular plane (Fig. 1.30).

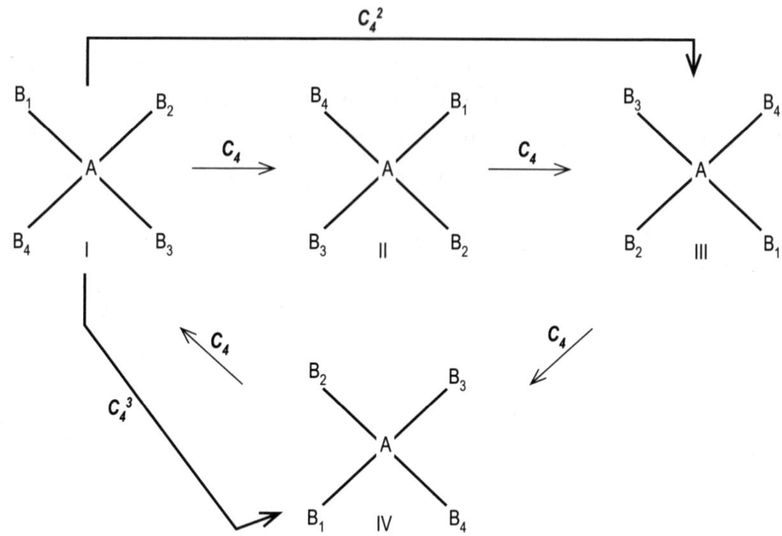

Fig. 1.30

When rotation is performed about the C_4 axis once through an angle of 90°, it leads to an indistinguishable configuration II. When rotation is performed further through the same angle three more times, it leads to the indistinguishable configurations III, IV and the original configuration I.

It is seen that when rotation about the C_4 axis is performed twice (C_4^2), it is equivalent to rotation through 180° (C_2). That is, $C_4^2 = C_2$.

As already pointed out, this molecule has four C_2 axes of symmetry. Out of these, two pass through A and bisect the line joining adjacent B atoms on both sides (Fig. 1.31).

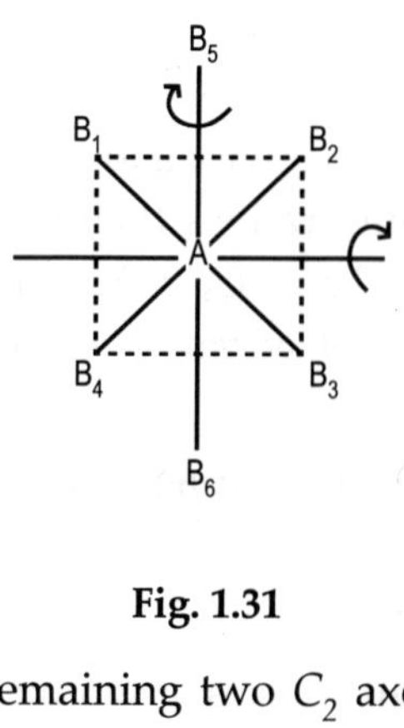

Fig. 1.31

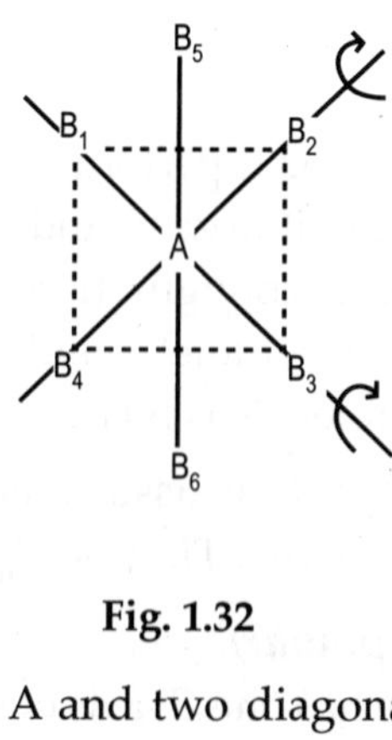

Fig. 1.32

The remaining two C_2 axes pass through A and two diagonally opposite B atoms (Fig. 1.32). All the four C_2 axes lie in the molecular plane and are perpendicular to the principal C_4 axis. An example of this type of molecule is $PtCl_4$.

Note: When a molecule has more than one C_n axis, the highest ordered axis is referred to as the principal axis of symmetry.

AB_4 (Tetrahedral)

This type of molecule has four C_3 axes of symmetry and three C_2 axes of symmetry. The C_3 axis is the imaginary line passing through A and one of the B atoms and the centre of the triangular plane whose corners are occupied by the other three B atoms (for example, the axis passing through B_1 and A is a C_3 axis and there are four such axes present in the molecule, Fig. 1.33).

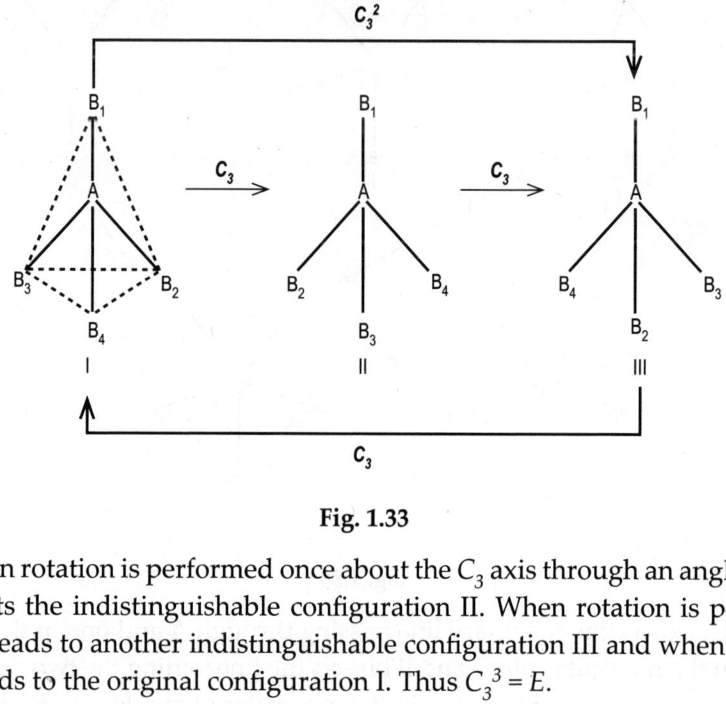

Fig. 1.33

When rotation is performed once about the C_3 axis through an angle of 120°, it presents the indistinguishable configuration II. When rotation is performed twice, it leads to another indistinguishable configuration III and when repeated thrice leads to the original configuration I. Thus $C_3^3 = E$.

The C_2 axis is the imaginary line which passes through A and bisects the line joining the front and back B atoms on both sides. Three C_2 axes are present in the molecule; two are shown (Fig. 1.34). An example, of this type of molecule is $[Ni(CO)_4]$.

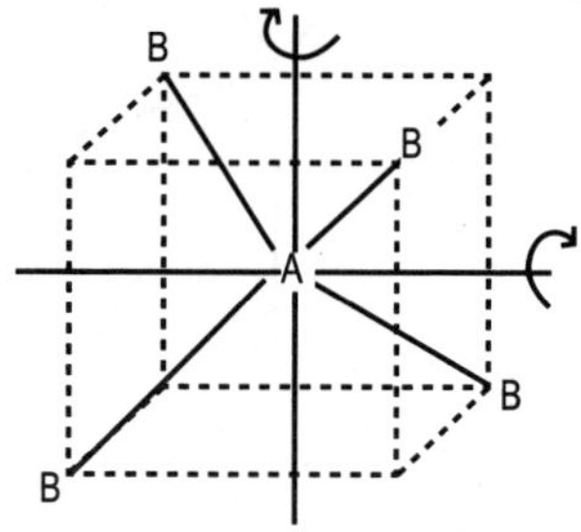

Fig. 1.34

***AB_5* (Pentagonal)**

This type of molecule has one C_5 axis of symmetry and five C_2 axes of symmetry. The C_5 axis is the imaginary line passing through A and is perpendicular to the molecular plane (Fig. 1.35). When rotation is performed once about the C_5 axis through an angle of 72°, it leads to an indistinguishable configuration II. When rotation is performed sequentially four times through the C_5 axis, it leads to the indistinguishable configurations III, IV, V and the original configuration I. Thus $C_5^5 = E$.

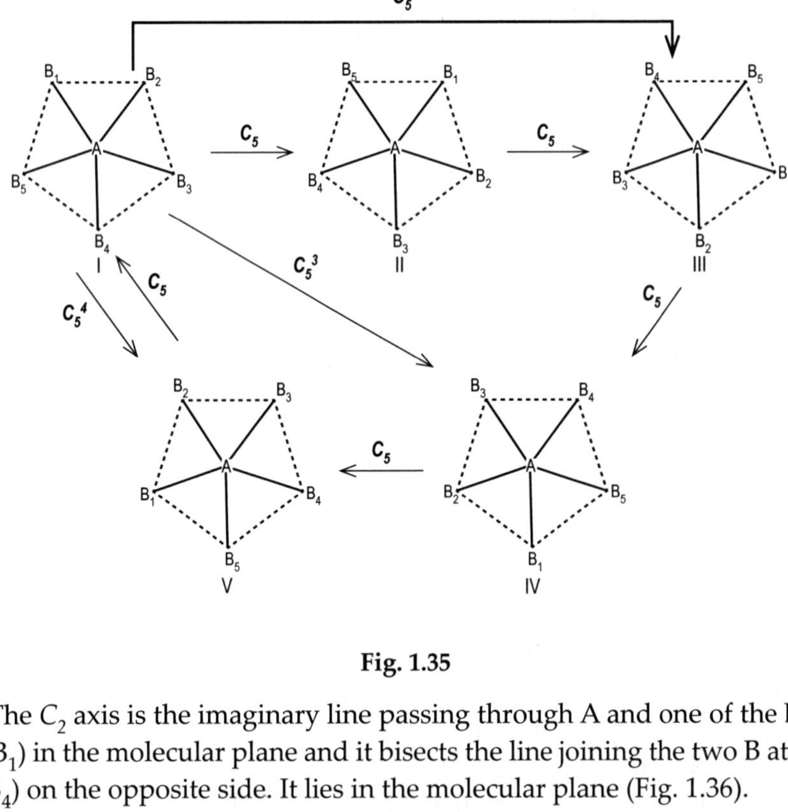

Fig. 1.35

The C_2 axis is the imaginary line passing through A and one of the B atoms (say, B_1) in the molecular plane and it bisects the line joining the two B atoms (B_3 and B_4) on the opposite side. It lies in the molecular plane (Fig. 1.36).

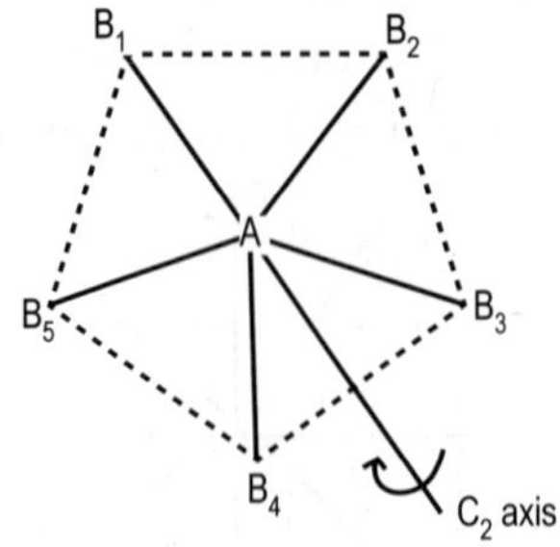

Fig. 1.36

There are five such axes present in this molecule and they are perpendicular to the principal C_5 axis.

AB_5 (Square pyramidal)

This type of molecule has a C_4 axis. This is an imaginary line that passes through A and the axial B atom (B_5 in Fig. 1.37). It is perpendicular to the square plane of the molecule. An example of this type of molcule is IF_5.

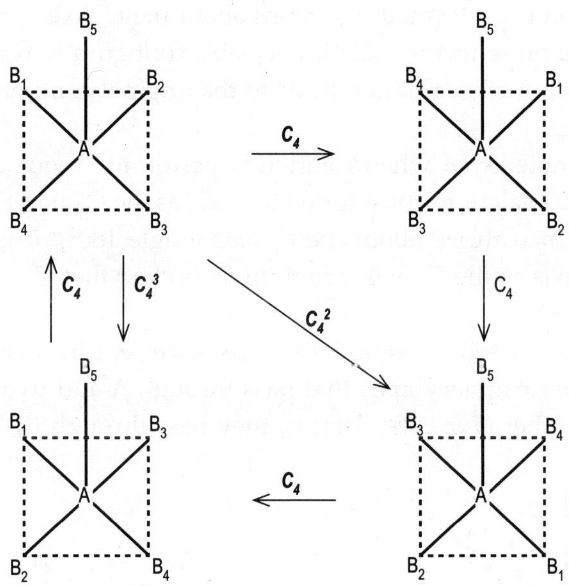

Fig. 1.37

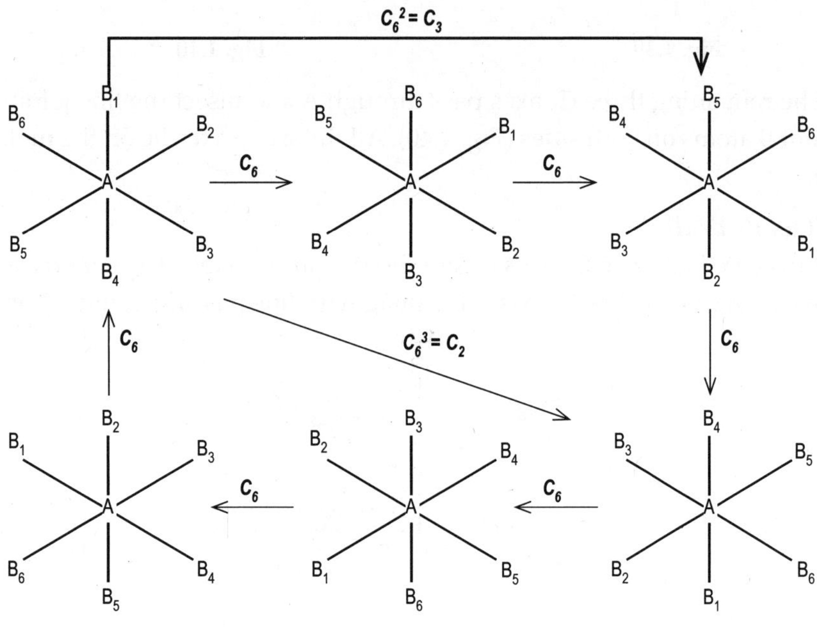

Fig. 1.38

AB_6 (Planar)

This type of molecule has one C_6 axis of symmetry and six C_2 axes of symmetry. The C_6 axis is the imaginary line passing through A and is perpendicular to the molecular plane (Fig.1.38).

When rotation is performed five times about the C_6 axis through an angle of 60° each time, it presents the indistinguishable configurations II, III, IV, V and VI respectively and further rotation leads to the original configuration. Thus, it is seen that $C_6^6 = E$.

It has to be noted that when rotation is performed twice through the C_6 axis (120°), it is equivalent to considering the axis as the C_3 axis. Similarly, when rotation is performed thrice about the C_6 axis (angle 180°), it is equivalent to considering the axis as the C_2 axis. From this it is clear that $C_6^2 = C_3^1$, $C_6^3 = C_2^1$, $C_6^4 = C_3^2$ and $C_6^6 = E$.

As already discussed, there are six C_2 axes present in this molecule. Out of these, three are imaginary lines that pass through A and two B atoms lying opposite to each other (Fig.1.39). That is, they pass through B_1–A–B_4, B_2–A–B_5 and B_3–A–B_6.

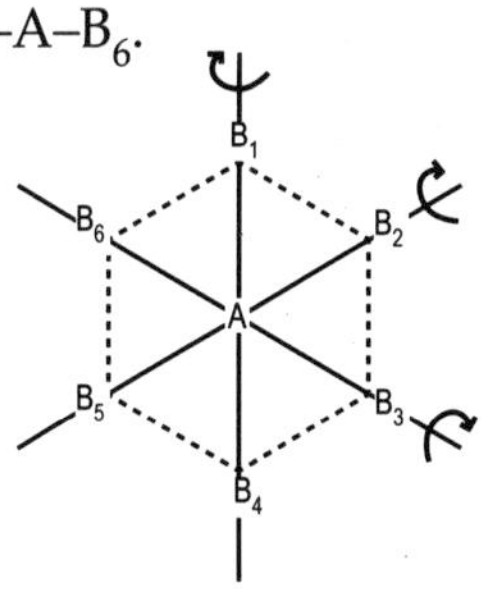

Fig. 1.39

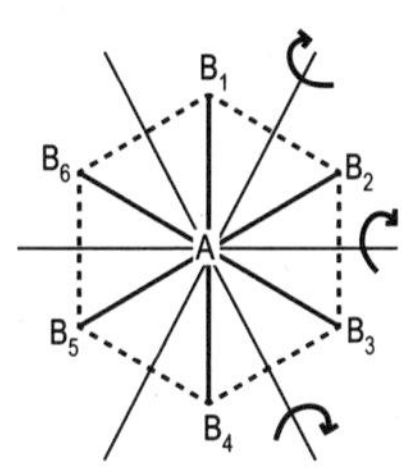

Fig. 1.40

The remaining three C_2 axes pass through A and bisect the line joining two adjacent B atoms on both sides (Fig. 1.40). All the six C_2 axes lie on the molecular plane.

AB_6 (Octahedral)

This molecule has three C_4 axes of symmetry, four C_3 axes of symmetry and six C_2 axes of symmetry. The C_4 axis is the imaginary line passing though A and two

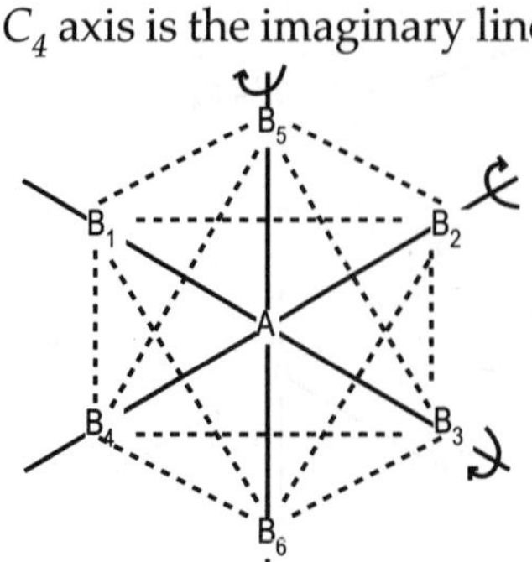

Fig. 1.41

diagonally opposite B atoms. There are three such axes present. They are, the lines passing through B_5–A–B_6, B_2–A–B_4 and B_1–A–B_3 (Fig. 1.41).

The C_3 axis is the imaginary line passing through A and the centres of the two triangles (two faces of the octahedron) opposite to each other. There are eight triangular faces in an octahedron and hence four C_3 axes are present in this molecule.

The C_2 axis is the imaginary line passing through A and bisecting the line joining the two adjacent B atoms on opposite sides. That is, it passes through A and two opposite edges of the octahedron. Since there are twelve edges in an octahedron, six C_2 axes exist in the molecule.

Improper axis of symmetry (S_n)

When a molecule is rotated about an axis through an angle of 360°/n, followed by reflection about a plane perpendicular to the originally chosen axis, if it presents an indistinguishable configuration, it is said to possess an improper axis of symmetry. This axis is represented by S_n. Molecules having an improper axis of symmetry can be classified into two types.

- Molecules having both, an axis of symmetry (C_n) and a perpendicular plane of symmetry (σ).
- Molecules having neither an axis of symmetry nor a plane of symmetry but having only an improper axis of symmetry (S_n).

An example for the first type is BCl_3 (Fig. 1.42). In this molecule, the C_3 axis is perpendicular to the molecular plane. Rotation through 120° (C_3) followed by reflection about the molecular plane gives an indistinguishable configuration. Hence, the molecule is said to possess an improper axis of symmetry (S_3). Hence, the C_3 axis is also the S_3 axis.

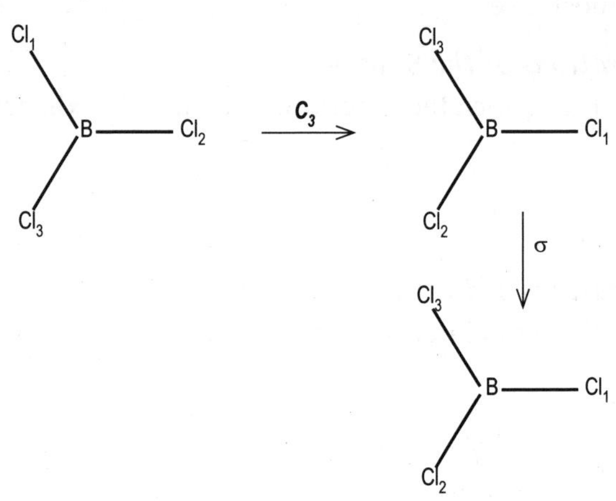

Fig. 1.42

An example for the second type is, the staggered conformation of ethane. Consider the end-on view (Newmann Projection) of the staggered conformation of ethane (Fig. 1.43).

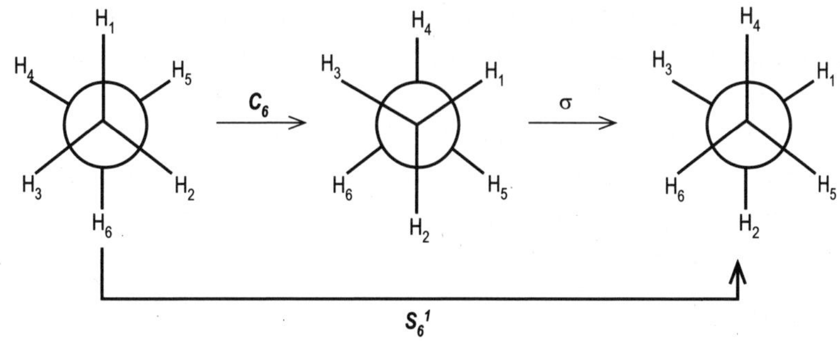

Fig. 1.43

Rotation about the C–C bond through an angle of 60° followed by reflection about a perpendicular plane at the centre of symmetry gives an indistinguishable configuration. Hence the C–C bond axis is the S_6 axis.

Note: No C_6 axis exists in this molecule, only the S_6 axis is present.

Associated operations of S_n

Rotation about an axis through an angle of $(360/n)°$ followed by reflection about a perpendicular plane can lead to an indistinguishable configuration but not the original configuration. However, by repeating the two processes many times, the original configuration can be arrived at. The number of times the process has to be repeated to get the original configuration is referred to as the *associated operations of the S_n axis*. The total number of associated operations of the S_n axis depends on whether n is even or odd. An assumption is made that C_n and σ_n commute which is true in most cases.

Associated operations of the S_2 axis

Mathematically, the associated operations of the S_2 axis are represented as follows.

$$S_2^1 = \sigma\, C_2^1$$
$$S_2^2 = \sigma_2\, C_2^2 = E$$

Associated operations of the S_4 axis

Mathematically, the associated operations of the S_4 axis are represented as follows.

$$S_4^1 = \sigma\, C_4^1$$
$$S_4^2 = \sigma^2\, C_4^2 = E\, C_2 = C_2$$
$$S_4^3 = \sigma^3\, C_4^3 = \sigma\, C_4^3$$
$$S_4^4 = \sigma^4\, C_4^4 = E \times E = E$$

Therefore, they can be listed as S_4^1, C_2, S_4^3 and E.

Associated operations of the S_6 axis

They are represented as,

$$S_6^1 = \sigma C_6^1$$
$$S_6^2 = \sigma^2 C_6^2 = E C_3 = C_3$$
$$S_6^3 = \sigma^3 C_6^3 = \sigma C_2 = i$$
$$S_6^4 = \sigma^4 C_6^4 = E C_3^2 = C_3^2$$
$$S_6^5 = \sigma^5 C_6^5 = \sigma C_6^5$$
$$S_6^6 = \sigma^6 C_6^6 = E \times E = E$$

Thus, an even-ordered improper axis of symmetry is associated with n number of operations where n is the order of the axis. It is interesting to note that the associated symmetry operations of the S_n axis (where n is even) include the associated symmetry operations of the $C_n/2$ axis. That is, the $C_n/2$ axis is also the S_n axis when n is even.

Associated operations of the S_3 axis

Mathematically, the associated operations are represented as,

$$S_3^1 = \sigma C_3^1$$
$$S_3^2 = \sigma^2 C_3^2 = C_3^2$$
$$S_3^3 = \sigma^3 C_3^3 = \sigma \times E = \sigma$$
$$S_3^4 = \sigma^4 C_3^4 = E C_3^1 = C_3^1$$
$$S_3^5 = \sigma^5 C_3^5 = \sigma C_3^2$$
$$S_3^6 = \sigma^6 C_3^6 = E \times E = E$$

Thus, the S_3 axis is associated with six symmetry operations.

Associated symmetry operations of the S_5 axis

The associated operations are represented as,

$$S_5^1 = \sigma C_5^1$$
$$S_5^2 = \sigma^2 C_5^2 = E \times C_5^2 = C_5^2$$
$$S_5^3 = \sigma^3 C_5^3 = \sigma C_5^3$$
$$S_5^4 = \sigma^4 C_5^4 = E \times C_5^2 = C_5^4$$
$$S_5^5 = \sigma^5 C_5^5 = \sigma \times E = \sigma$$
$$S_5^6 = \sigma^6 C_5^6 = E \times E \times C_5^1 = C_5^1$$
$$S_5^7 = \sigma^7 C_5^7 = \sigma \times E \times C_5^2$$
$$S_5^8 = \sigma^8 C_5^8 = E \times E \times C_5^3 = C_5^3$$
$$S_5^9 = \sigma^9 C_5^9 = \sigma \times E \times C_5^4$$
$$S_5^{10} = \sigma^{10} C_5^{10} = E \times E \times E = E$$

Thus, the S_5 axis is associated with 10 symmetry operations. In general, the S_n axis of odd order is associated with $2n$ number of symmetry operations. The associated operations of the odd ordered S_n axis include the associated operations of the C_n axis. This indicates that the C_n axis of odd order is also the S_n axis of the same order.

Horizontal planes (σ_h), vertical planes (σ_v) and dihedral planes (σ_d)

The various symmetry planes present in molecules of different shapes have already been discussed in detail. These planes can be classified into three types—horizontal planes (σ_h), vertical planes (σ_v) and dihedral planes (σ_d).

Horizontal planes (σ_h)

Horizontal planes are those planes that are perpendicular to the principal axis of the molecule. For example, in the BCl_3 molecule, the C_3 axis is the principal axis. It passes through boron and is perpendicular to the molecular plane. Hence, the molecular plane becomes the horizontal plane as it is perpendicular to the principal axis.

Vertical planes (σ_v)

Planes containing the principal axis are referred to as vertical planes. For example, in the water molecule, the principal axis is the C_2 axis passing through oxygen. Both the molecular plane and the other plane passing through oxygen and bisecting the line joining the two hydrogen atoms contain the C_2 axis and are thus referred to as vertical planes.

Dihedral planes (σ_h)

Planes bisecting two C_2 axes are referred to as dihedral planes. For example, the planes that exist in the staggered conformation of ethane are dihedral planes.

1.4 PRODUCT OF SYMMETRY OPERATIONS

The product of any two symmetry operations is also a symmetry operation. The product depends on the order in which the symmetry operations are carried out. Generally, the order in which symmetry operations are performed is represented from right to left.

For example, if the symmetry operation A is carried out first followed by B, it is represented as $B \times A$.

1.4.1 Determination of the Product of Symmetry Operations

The product of symmetry operations can be determined either by choosing a point with its respective coordinates (Method 1) or an appropriate molecule (Method 2) or through a matrix (Method 3). All these methods can also be extended to determine the product of more than two symmetry operations.

Method 1

Let a symmetry operation A be performed on a point with its coordinates as x_1, y_1, z_1. Let the transformed coordinates be x_2, y_2, z_2. Now, if a second symmetry operation B is performed on x_2, y_2, z_2, let the resultant coordinates be x_3, y_3, z_3. A single symmetry operation C has to be identified which can directly transform x_1,y_1,z_1 to x_3,y_3,z_3 which is the product of the symmetry operations A followed by B performed on the point as shown below.

$$(x_1,y_1,z_1) \xrightarrow{A} (x_2,y_2,z_2) \xrightarrow{B} (x_3,y_3,z_3)$$

$$(x_1,y_1,z_1) \xrightarrow{C} (x_3,y_3,z_3)$$

For example, consider a rotation by 180° about the x axis followed by a rotation by 180° about the z axis.

$$(x_1,y_1,z_1) \xrightarrow{C_{2(x)}} (x_1,-y_1,-z_1) \xrightarrow{C_{2(z)}} (-x_1,y_1,-z_1)$$

$$(x_1,y_1,z_1) \xrightarrow{C_{2(y)}} (-x_1,y_1,-z_1)$$

$$\therefore C_{2(z)} \times C_{2(x)} = C_{2(y)}$$

The effect is the same as a single rotation by 180° about the y axis.

Limitation: The limitation of this method is that it cannot be applied to products involving higher ordered proper axis of symmetry (C_n).

Method 2

A molecule that has both the symmetry operations for which the product has to be determined, is chosen. The desired symmetry operations are performed one after another in the respective order and the new configuration is obtained for the molecule. Now, a single symmetry operation which can directly transform the molecule from configuration I to configuration III has to be identified which is the product of the two symmetry operations. For example, to find the product of $C_{3(z)}$ followed by σ_{v1}. Consider the ammonia molecule since it contains both C_3 and σ_v.

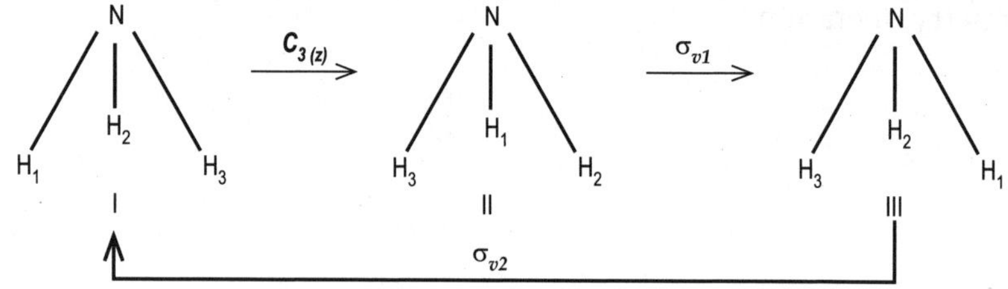

Fig. 1.44

σ_{v1} is the plane passing through nitrogen and the hydrogen on the left and σ_{v2} is the plane passing through nitrogen and the hydrogen at the back.

$$\sigma_{v1} \times C_{3(2)} = \sigma_{v2}$$

Method 3

The matrix method is a complete method and is dealt with later (Section 3.17.5).

1.4.2 Commutative Symmetry Operations

Two symmetry operations *A* and *B* are said to be commutative when the product of these two symmetry operations remains the same irrespective of the order in which they are carried out. That is,

$$A \times B = B \times A = C$$

To find out whether σ_{xz} and σ_{xy} are commutative, the two binary products $\sigma_{xz} \times \sigma_{xy}$ and $\sigma_{xy} \times \sigma_{xz}$ are determined. If the products are the same, then these two operations are said to be commutative.

$$(x_1, y_1, z_1) \xrightarrow{\sigma_{xy}} (x_1, y_1, -z_1) \xrightarrow{\sigma_{xz}} (x_1, -y_1, -z$$

$$(x_1, y_1, z_1) \xrightarrow{C_{2(x)}} (x_1, -y_1, -z_1)$$

$$\therefore \quad \sigma_{xz} \times \sigma_{xy} = C_{2(x)}$$

$$(x_1, y_1, z_1) \xrightarrow{\sigma_{xz}} (x_1, -y_1, z_1) \xrightarrow{\sigma_{xy}} (x_1, -y_1, -z_1)$$

$$(x_1, y_1, z_1) \xrightarrow{C_{2(x)}} (x_1, -y_1, -z_1)$$

Therefore, $\sigma_{xy} \times \sigma_{xz} = C_{2(x)}$

Since $\sigma_{xz} \times \sigma_{xy} = \sigma_{xy} \times \sigma_{xz} = C_{2(x)}$, these two operations are commutative.

This chapter highlights the definitions for symmetry elements and symmetry operations, how to identify the symmetry elements present in molecules of different geometries and the methods used for the determination of the products of symmetry operations.

2

Symmetry Elements and Point Groups

2.1 Group

This chapter takes us from the identification of symmetry elements that molecules possess, to the classification of molecules into groups. A group is defined as any set or collection of elements which obey a certain set of mathematical rules and are related to each other by certain rules. The elements could be numbers, matrices, vectors, roots of an equation or symmetry elements. For example,

- All positive and negative whole numbers together with zero form a group under the combining operation of addition.
- Elements 1, –1, i, –i (where $i = \sqrt{-1}$) form a group under multiplication.
- All powers of two ....2^{-2}, 2^{-1}, 2^0, 2^1, 2^2.... form an infinite group under multiplication.

2.2 Mathematical Rules for the Formation of a Group

The elements of any group must obey the following rules.

- The combination of any two members of a group and the square of every member of a group should also be a member of that group (this is known as the *closure property*). Let A and B be the members of a group under the multiplication operation. If A multiplied by B gives C, A multiplied by A gives D, and B multiplied by B gives G, then C, D and G should also be members of the group. (C, D and G can also be the same as A or B.)
- The group must contain one and only one identity element (*E*) which has

the property $AE = EA = A$ for all elements of the group.

- The associative law of combination should hold good.
  That is, A (B C) = (A B) C
- The inverse of every element should also be a member of the group. The inverse of any element X is denoted by X^{-1} which when combined with the element itself results in the identity element. It can be represented by the following equation.

$$X^{-1}X = X X^{-1} = E$$

Let the inverse of σ be σ^{-1}. Therefore, $\sigma \sigma^{-1} = E$ according to the definition. Since it has already been proved that $\sigma^2 = E$, $\sigma^{-1} = \sigma$.
Similarly, we can prove that the inverse of ***i*** is ***i*** itself, and C_2 is C_2 itself. But the inverse of C_3 is C_3^2 since $C_3^3 = E$. In general, the inverse of any C_n is $(C_n)^{n-1}$ and the inverse of any S_n is $(S_n)^{n-1}$ when n is even and $(S_n)^{2n-1}$ when n is odd.

2.3 Definition and Classification of Point Groups

A point group is a collection of the symmetry elements present in a molecule that obeys the mathematical rules for the formation of a group. During the application of various symmetry operations in a molecule, the centre of mass does not shift and therefore all the axes, points and planes of symmetry of the molecule intersect at one common invariant point (center of mass) and hence the term *point group* is used. The point groups are represented by *Schoenflies symbols*. Table 2.1 summarises the various point groups and their corresponding symmetry elements.

Table 2.1 Point groups and their symmetry elements

Point group	Symmetry elements
C_1	E
C_s	E, σ
C_i	E, i
C_2	E, C_2
C_3	E, C_3^1, C_3^2
C_4	E, C_4^1, $C_4^2 = C_2$, C_4^3
C_5	E, C_5^1, C_5^2, C_5^3, C_5^4
C_6	E, C_6^1, $C_6^2 = C_3$, $C_6^3 = C_2$, $C_6^4 = C_3^2$, C_6^5
C_{2v}	E, C_2, σ_{v1}, σ_{v2}
C_{3v}	E, C_3^1, C_3^2, σ_{v1}, σ_{v2}, σ_{v3}
C_{4v}	E, C_4^1, $C_4^2 = C_2$, C_4^3, σ_{v1}, σ_{v2}, σ_{v3}, σ_{v4}

C_{5v}	$E, C_5^1, C_5^2, C_5^3, C_5^4, 5\sigma_v$
C_{6v}	$E, C_6^1, C_6^2 = C_3, C_6^3 = C_2, C_6^4 = C_3^2, C_6^5, 6\sigma_v$
$C_{\infty v}$	E, C_∞, infinite number of σ_v
C_{2h}	E, C_2, σ_h, i
C_{3h}	$E, C_3^1, C_3^2, \sigma_h, S_3^1, S_3^5$
C_{4h}	$E, C_4^1, C_4^2 = C_2, C_4^3, \sigma_h, S_4^1, S_4^3$
C_{5h}	$E, C_5^1, C_5^2, C_5^3, C_5^4, \sigma_h, S_5^1, S_5^3, S_5^7, S_5^9$
C_{6h}	$E, C_6^1, C_6^2 = C_3, C_6^3 = C_2, C_6^4 = C_3^2, C_6^5, \sigma_h, S_3^1, S_3^5, i, S_6^1, S_6^5$
D_2	$E, C_2, 2 \perp C_2$
D_3	$E, C_3^1, C_3^2, 3 \perp C_2$
D_{2h}	$E, C_2, 2 \perp C_2, 2\sigma_h, 2\sigma_v, S_2 = i$
D_{3h}	$E, C_3^1, C_3^2, 3C_2 \perp C_3, \sigma_h, S_3^1, S_3^5, 3\sigma_v$
D_{4h}	$E, C_4^1, C_4^2 = C_2, C_4^3, 4C_2 \perp C_4, \sigma_h, 4\sigma_v, S_4^1, S_4^3, i$
D_{5h}	$E, C_5^1, C_5^2, C_5^3, C_5^4, 5C_2 \perp C_5, \sigma_h, 5\sigma_v, S_5^1, S_5^3, S_5^7, S_5^9$
D_{6h}	$E, C_6^1, C_6^2 = C_3, C_6^3 = C_2, C_6^4 = C_3^2, C_6^5, 6C_2 \perp C_6, \sigma_h, 6\sigma_v$ $S_3^1, S_3^5, i, S_6^1, S_6^5$
$D_{\infty h}$	E, C_∞, infinite number of C_2, σ_h infinite number of σ, i
D_{2d}	$E, 2\sigma_d$, 3 mutually $\perp C_2, S_4^1, S_4^3$
D_{3d}	$E, C_3^1, C_3^2, 3\ C_2 \perp C_3, 3\sigma_d, S_6^1, S_6^5, i$
D_{4d}	$E, C_4^1, C_4^2 = C_2, C_4^3, 4\ C_2 \perp C_4, 4\sigma_d, S_8^1, S_8^3, S_8^5, S_8^7$
D_{5d}	$E, C_5^1, C_5^2, C_5^3, C_5^4, 5C_2 \perp C_5, 5\sigma_d, S_{10}^1, S_{10}^3, S_{10}^7, S_{10}^9, i$
D_{6d}	$E, C_6^1, C_6^2 = C_3, C_6^3 = C_2, C_6^4 = C_3^2, C_6^5, 6C_2 \perp C_6, 6\sigma_d,$ $S_4^1, S_4^3, S_6^1, S_6^5, S_{12}^1, S_{12}^{11}$
S_4	E, C_2, S_4^1, S_4^3
S_6	$E, C_3^1, C_3^2, S_6^1, S_6^5, i$
T_d	$E, 8C_3, 3C_2, 6S_4, 6\sigma_d$
O_h	$E, 6C_4, 3C_4^2 = C_2, 8C_3, 6C_2, 3\sigma_h, 6\sigma_d, I, 8S_6, 6S_4$
I_h	$E, 12C_5^1, 12C_5^2, 20C_3, 15C_2, 15\sigma_d, 20S_6, 12S_{10}^1, i, 12S_{10}^3$

The point groups C_1, C_s and C_i are generally referred to as low symmetry point groups as they have either one or two symmetry elements.

The C_n point groups have the C_n axis as the only symmetry element. However, n number of associated symmetry operations of the C_n axis are present in the C_n point groups. The C_{nh} point groups have an additional σ_h along with the C_n axis.

The C_n axis can generate n number of associated operations and the combination of C_n and σ_h can generate n number of symmetry elements corresponding to the S_n axis thus accounting for $2n$ symmetry operations.

$$E, C_n^1, C_n^2, \text{---------} C_n^{n-1}, \sigma_h, S_n^1, S_n^3, \text{---------} S_n^{n-1} \text{ or } S_n^{2n-1}$$

The C_{nv} point groups contain the C_n axis and $n\,\sigma_v$. The C_n axis can generate n symmetry elements and an additional n number of vertical planes thus accounting for $2n$ symmetry elements (E, C_n^1, C_n^2--------- $C_n^{n-1}, n\sigma_v$).

D_n point groups contain, in addition to the C_n axis, a set of nC_2 axes perpendicular to it. n number of associated operations of the C_n axis and n number of C_2 axes perpendicular to the C_n axis together account for the $2n$ symmetry elements of this group.

D_{nh} point groups have a σ_h in addition to a set of D_n group elements; that is, C_n and n perpendicular C_2 axes. The presence of σ_h and nC_2 axes perpendicular to C_n automatically implies the presence of $n\,\sigma_v$. C_n axis and σ_h can generate n number of S_n symmetry elements. Thus there are $4n$ number of symmetry elements present in this group.

D_{nd} point groups have the C_n axis, nC_2 axes perpendicular to it and n number of vertical planes. Such groups are additionally associated with n number of symmetry operations of the S_{2n} axis. Thus, a total of $4n$ symmetry elements are present in the point group.

Symmetry axes of special point groups like T_d and O_h are illustrated in the following figures

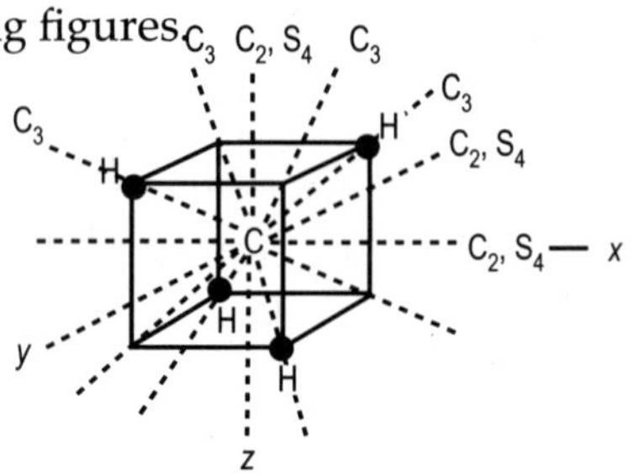

T_d point group

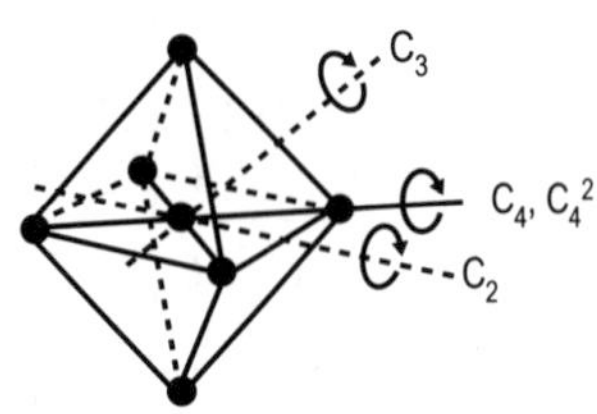

O_h point group

With experience, one learns to recognise the point group to which a molecule belongs simply by analogy with some other known molecule. However, until one builds up a memory bank of the point groups of representative molecules, it is best to use the systematic method. A scheme which will enable the reader to systematically deduce this, is shown in the flow diagram (Fig. 2.1). It can be understood by studying the flow chart that all the symmetry elements present in a molecule need not be determined to assign the point group. To illustrate how it works, some general examples are worked out (Tables 2.2A to 2.2I) followed by some specific examples in Table 2.3.

Fig. 2.1 Flow chart for determination of point group of a molecule

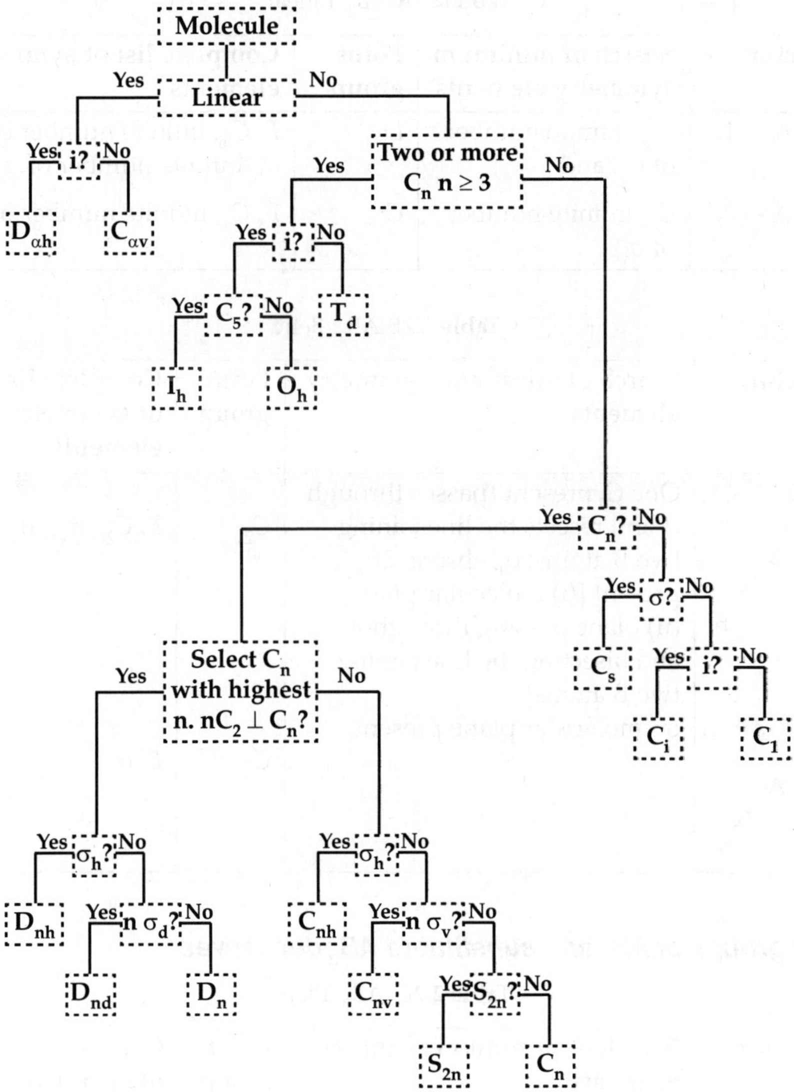

2.3.1 Point Groups of AB_n and Substituted Derivatives

The Tables 2.2A to 2.2I indicate how point groups can be arrived at for molecules of different geometries – having the general structures AB_2, AB_3, AB_4, AB_5 and AB_6 – by searching for the symmetry elements as outlined in the flowchart (Fig 2.1). The tables also provide information regarding the variation in symmetry elements and point groups that is brought about when the B atoms are gradually replaced by X or Y atoms. The complete list of symmetry elements that is present in each type of molecule has been provided.

Point groups of AB_2 and substituted AB_2 derivatives

Table 2.2A AB_2: Linear

Structure	Search of minimum symmetry elements	Point group	Complete list of symmetry elements
B — A — B	C_∞, infinite number of C_2, and i	$D_{\infty h}$	E, C_∞, infinite number of C_2, σ_h, infinite number of σ and i
B — A — X	C_∞, infinite number of σ_v	$C_{\infty v}$	E, C_∞, infinite number of σ_v

Table 2.2B AB_2: Bent

Structure	Search of minimum symmetry elements	Point group	Complete list of symmetry elements
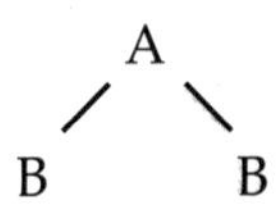	One C_2 present (passes through A and bisects the line joining two B atoms, σ_h absent, $2\sigma_v$ present [(i) molecular plane (ii) plane passing through A and bisecting the line joining two B atoms].	C_{2v}	E, C_2, σ_{v1}, σ_{v2}
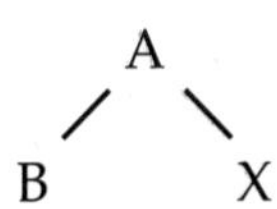	σ - molecular plane present.	C_s	E, σ

Point groups of AB_3 and substituted AB_3 derivatives

Table 2.2C AB_3: Planar

Structure	Search of minimum symmetry elements	Point group	Complete list of symmetry elements
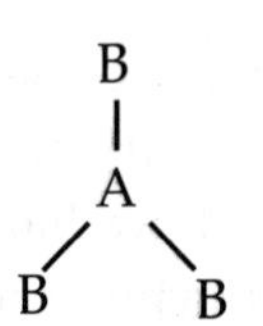	C_3 axis present (passes through A and is perpendicular to the molecular plane), $3C_2 \perp$ to C_3 present (each C_2 passes through A and one of the B atoms and bisects the line joining the other two B atoms), σ_h present (molecular plane).	D_{3h}	E, C_3^1, C_3^2, $3\,C_2 \perp C_3$, σ_h, S_3^1, S_3^5, $3\sigma_v$

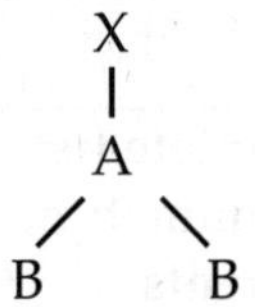	One C_2 axis present (passes through X and A), σ_h absent, $2\sigma_v$ present (one molecular plane and another plane passing through X, A and bisecting the line joining two B atoms).	C_{2v}	$E, C_2, \sigma_{v1}, \sigma_{v2}$
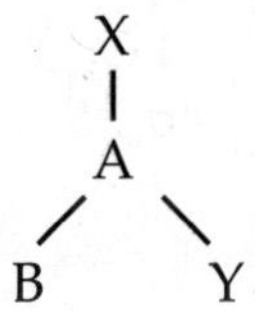	σ present (molecular plane).	C_s	E, σ

Table 2.2D AB_3: Pyramidal

Structure	Search of minimum symmetry elements	Point group	Complete list of symmetry elements
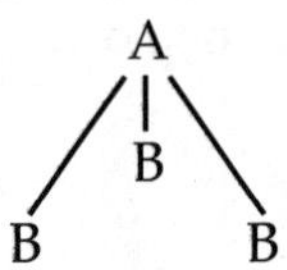	One C_3 axis present which passes through A and the centre of the triangle having B atoms at the corners, σ_h absent, $3\sigma_v$ present (each passes through A and one of the B atoms and bisects the line joining the other two B atoms).	C_{3v}	$E, C_3^1, C_3^2, \sigma_{v1}, \sigma_{v2}, \sigma_{v3}$
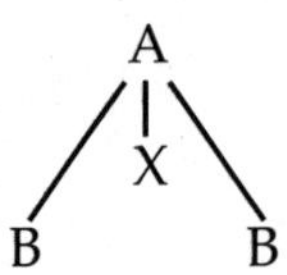	σ (passes through A and X atoms and bisects the line joining the two B atoms).	C_s	E, σ
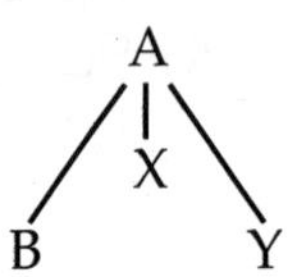	C_n axis absent, σ absent.	C_1	E

Point groups of AB_4 and substituted AB_4 derivatives

Table 2.2E AB_4: Square planar

Structure	Search of minimum symmetry elements	Point group	Complete list of symmetry elements
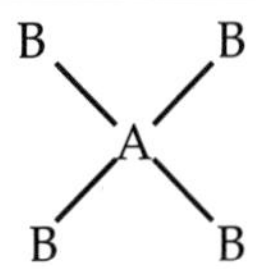	C_4 axis present (passing through A and ⊥ to the molecular plane), four C_2 axes also present (2 C_2 axes pass through A and diagonally opposite B atoms, other two C_2 axes pass through A and bisect the line joining the adjacent B atoms on both sides), σ_h present (molecular plane), $4\sigma_v$ present (two σ_v pass through A and bisect the line joining the adjacent B atoms on both sides. Two other σ_v pass through A and the diagonally opposite B atoms.	D_{4h}	$E, C_4^1, C_4^2 = C_2, C_4^3, 4\,C_2 \perp C_4, \sigma_h, 4\sigma_v, S_4^1, S_4^3, i$
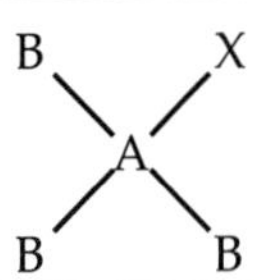	One C_2 axis present (passes through X , A and B), σ_h absent, $2\sigma_v$ present (molecular plane and plane passing through X, A and B).	C_{2v}	$E, C_2, \sigma_{v1}, \sigma_{v2}$
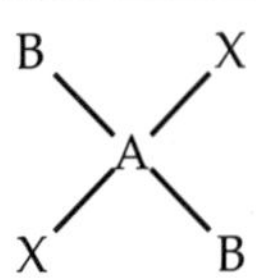	$3C_2$ axes present (principal C_2 axis passes through Pt and is perpendicular to the molecular plane, one C_2 axis passes through Pt and two B atoms and other axis passes through Pt and two X atoms, σ_h present (molecular plane).	D_{2h}	$E, C_2, 2 \perp C_2, \sigma_h, 2\sigma_v, S_2 = i$
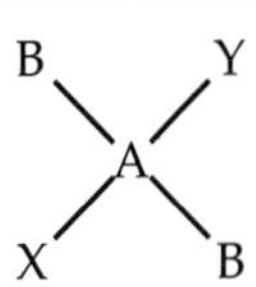	One C_2 axis present (passes through Y, A and X), σ_h absent, $2\sigma_v$ present (molecular plane and plane passing through X, A and Y).	C_{2v}	$E, C_2, \sigma_{v1}, \sigma_{v2}$
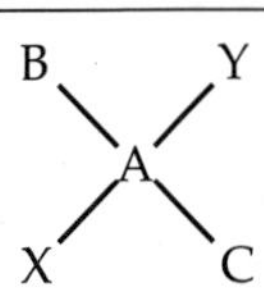	Only σ (molecular plane) present.	C_s	E, σ

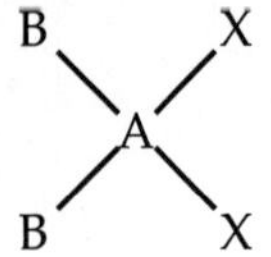	One C_2 axis present (passes through A and bisects the line joining two B atoms on one side and two X atoms on the other side), σ_h absent, $2\sigma_v$ present (molecular plane and plane passing through A and bisecting the lines joining two B atoms on one side and two X atoms on other side).	C_{2v}	$E, C_2, \sigma_{v1}, \sigma_{v2}$
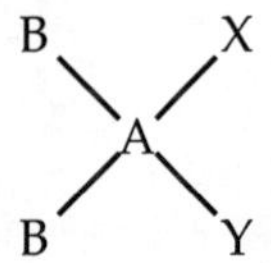	Only σ (molecular plane) present.	C_s	E, σ
	Only σ (molecular plane) present.	C_s	E, σ

Table 2.2F AB_4: Tetrahedral

Structure	Search of minimum symmetry elements	Point group	Complete list of symmetry elements
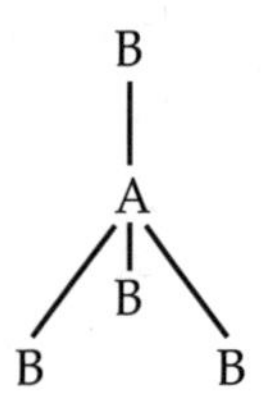	$4C_3$ axes present (pass through B, A and the centre of the triangle having B atoms at the corners), $3C_2$ axes present (pass through A and bisect line joining two front B atoms and two back B atoms), $6\sigma_d$ present (pass through A and two B atoms).	T_d	$E, 8C_3, 3C_2, 6S_4, 6\sigma_d$
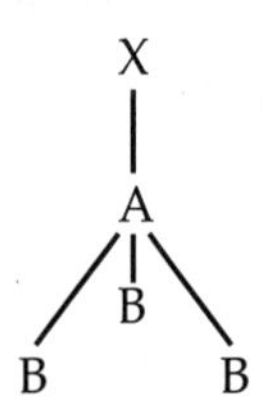	C_3 axis present (passes through X, A and the centre of the triangle having B atoms at the corners), σ_h absent, $3\sigma_v$ present (each passes through X, A and one of the B atoms and bisects the line joining the other two B atoms).	C_{3v}	$E, C_3^1, C_3^2, \sigma_{v1}, \sigma_{v2}, \sigma_{v3}$
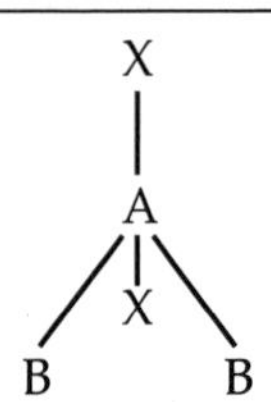	One C_2 axis present (passes through A and bisects two B atoms and two X atoms), $2\sigma_v$ present (one plane passes through A and two B atoms and the other passes through A and two X atoms).	C_{2v}	$E, C_2, \sigma_{v1}, \sigma_{v2}$

	One σ present (passes through X, A and Y and bisects the line joining the two B atoms).	C_s	E, σ
	C_n axis absent, σ absent.	C_1	E

Point groups of AB_5 and substituted AB_5 derivatives

Table 2.2G AB_5: Trigonal bipyramidal

Structure	Search of minimum symmetry elements	Point group	Complete list of symmetry elements
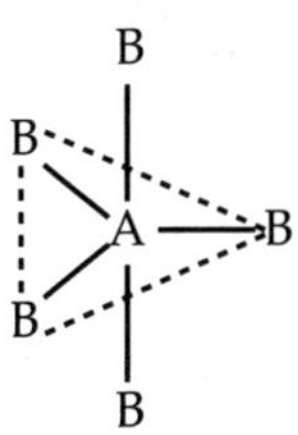	C_3 axis present (passes through A and two axial B atoms and is perpendicular to the plane having A and the three equatorial B atoms), $3C_2 \perp$ to C_3 present (each C_2 passes through A and one of the equatorial B atoms and bisects the line joining the other two B atoms), σ_h present (plane passes through A and three equatorial B atoms).	D_{3h}	$E, C_3^1, C_3^2, 3\ C_2 \perp C_3, \sigma_h, S_3^1, S_3^5, 3\sigma_v$
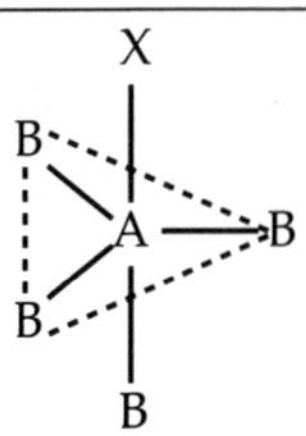	C_3 axis present (passes through X, A and axial B atom), σ_h absent, $3\sigma_v$ present (each passes through X, A, one equatorial B atom and one axial B atom).	C_{3v}	$E, C_3^1, C_3^2, \sigma_{v1}, \sigma_{v2}, \sigma_{v3}$

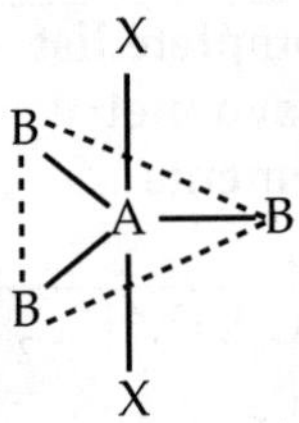	C_3 axis present (passes through A and two axial X atoms and is perpendicular to the plane having A and three B atoms), $3C_2 \perp$ to C_3 present (each C_2 passes through A and one of the B atoms and bisects the line joining the other two B atoms), σ_h present (plane passing through A and the equatorial B atoms).	D_{3h}	$E, C_3^1, C_3^2, 3\ C_2 \perp C_3, \sigma_h, S_3^1, S_3^5, 3\sigma_v$
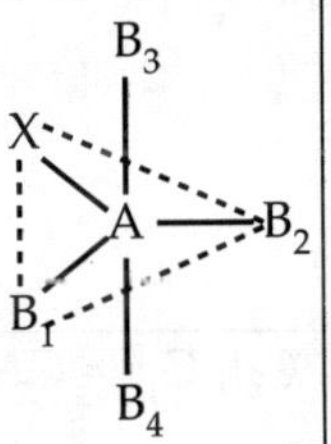	One C_2 axis present (passes through X, A and bisects the line joining the two B atoms), σ_h absent, $2\sigma_v$ present (equatorial plane and plane passing through X, A, B_3 and B_4, and bisecting the line joining other two B atoms).	C_{2v}	$E, C_2, \sigma_{v1}, \sigma_{v2}$
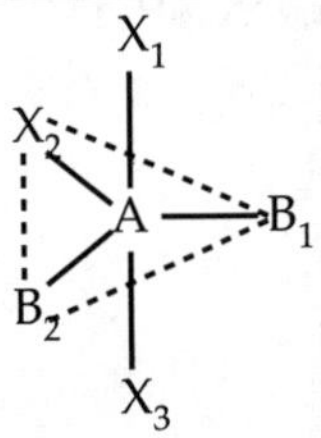	One C_2 axis present (passes through X_2, A and bisects the line joining B_1 and B_2), σ_h absent, $2\sigma_v$ present (plane having X_2, A, B_1 and B_2, and plane passing through X_2, A, X_1 and X_3).	C_{2v}	$E, C_2, \sigma_{v1}, \sigma_{v2}$
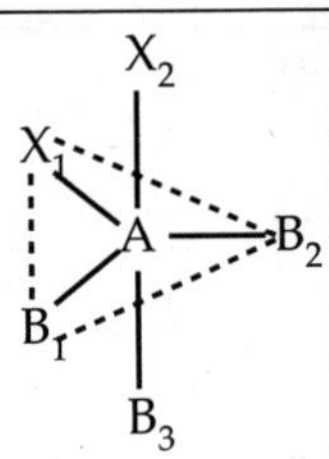	One σ present (passes through X_1, X_2, A and B_3).	C_s	E, σ
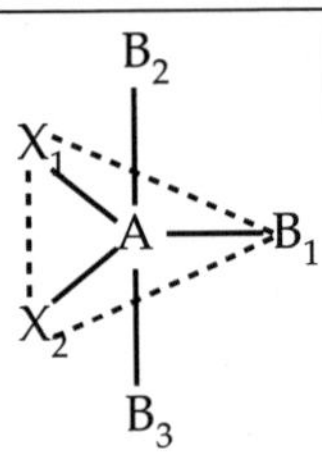	One C_2 axis present (passes through B_1 and A), σ_h absent, $2\sigma_v$ present (plane having B_1, X_1, A and X_2, and plane passing through B_1, A, B_2 and B_3 and bisecting the line joining two X atoms).	C_{2v}	$E, C_2, \sigma_{v1}, \sigma_{v2}$
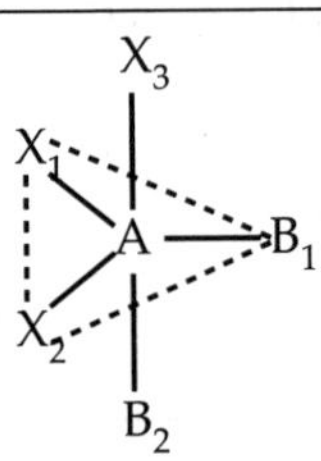	One σ present (passes through X_3, B_1, A and B_2).	C_s	E, σ

Table 2.2H AB_5: Square pyramidal

Structure	Search of minimum symmetry elements	Point group	Complete list of symmetry elements
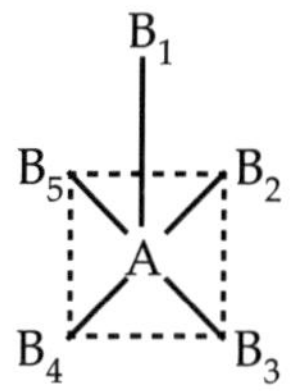	One C_4 present (passes through axial B_1 and A), perpendicular C_2 absent, σ_h absent, $4\sigma_v$ present (two mutually perpendicular σ_v pass through B_1, A and bisect the line joining adjacent B atoms, two mutually perpendicular planes of symmetry pass through B_1, A and diagonally opposite equatorial B atoms).	C_{4v}	$E, C_4^1, C_4^2 = C_2, C_4^3, \sigma_{v1}, \sigma_{v2}, \sigma_{v3}, \sigma_{v4}$
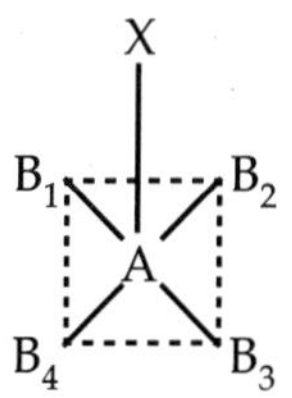	One C_4 present (passes through X and A), perpendicular C_2 absent, σ_h absent, $4\sigma_v$ present (two mutually perpendicular σ_v pass through X, A and bisect the line joining adjacent B atoms, two mutually perpendicular planes of symmetry pass through X, A and diagonally opposite B atoms).	C_{4v}	$E, C_4^1, C_4^2 = C_2, C_4^3, \sigma_{v1}, \sigma_{v2}, \sigma_{v3}, \sigma_{v4}$
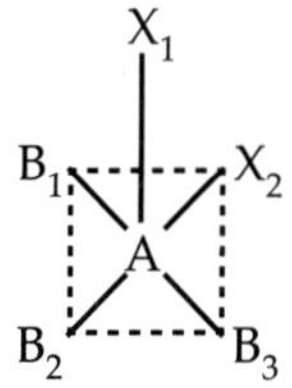	One σ present (passes through X_1, X_2, A and B_2).	C_s	E, σ
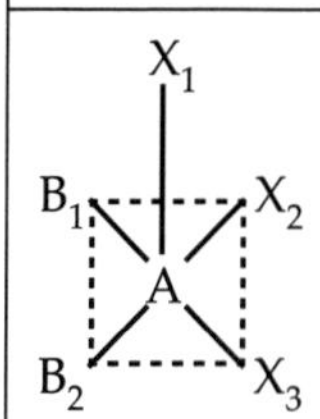	One σ present (passes through X_1 and A and bisects the line joining the two B atoms and the two X atoms in the plane).	C_s	E, σ
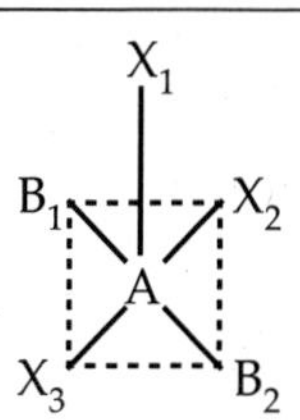	One C_2 present (passes through A and axial X), $2\sigma_v$ present (one σ_v passes through X_1, X_3, A and X_2, the other passes through B_1, B_2, A and X_1.	C_{2v}	$E, C_2, \sigma_{v1}, \sigma_{v2}$

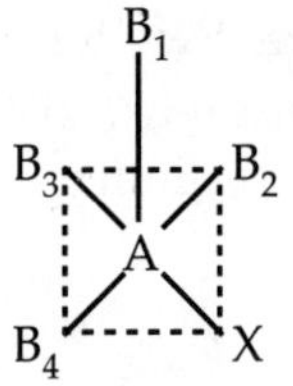	One σ present (passes through B_1, B_3, X, A).	C_s	E, σ

Point groups of AB_6 and substituted AB_6 derivatives

Table 2.2I AB_6: Octahedral

Structure	Search of minimum symmetry elements	Point group	Complete list of symmetry elements
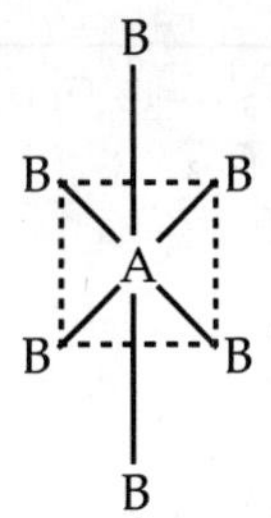	$3C_4$ axis present (each C_4 passes through A and two diagonally opposite B atoms), $4C_3$ present (pass through A and opposite faces), $6C_2$ present (pass through A and opposite edges), $3\sigma_h$ present (square planes having A and four B atoms), $6\sigma_v$ present (pass through opposite faces).	O_h	$E, 6C_4, 3C_4^2 = C_2, 8C_3, 6\,C_2, 3\sigma_h, 6\sigma_d$, i, $8S_6, 6S_4$
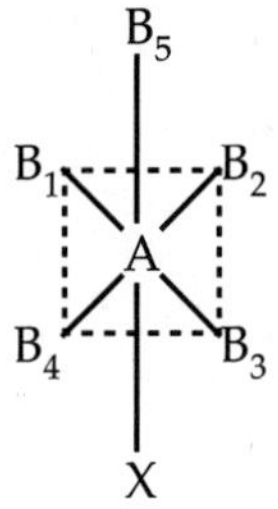	One C_4 present (passes through X, A and axial B atom), perpendicular C_2 absent, σ_h absent, $4\sigma_v$ present (two mutually perpendicular σ_v pass through X, A and axial B atom and bisect the line joining the two B atoms on both sides, and two mutually perpendicular planes of symmetry pass through B_1, B_5, A, X, B_3 and X, B_2, A, B_4, B_5).	C_{4v}	$E, C_4^1, C_4^2 = C_2, C_4^3, \sigma_{v1}, \sigma_{v2}, \sigma_{v3}, \sigma_{v4}$
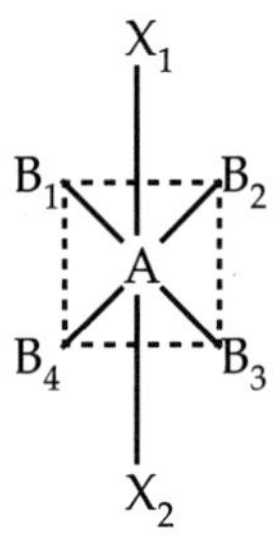	One C_4 present (passes through X_1, A, X_2), $4C_2 \perp C_4$ present (two C_2 axes pass through A and diagonally opposite B atoms, two C_2 axes pass through the centers of line, joining adjacent B atoms), σ_h is present (square plane passing through A and four B atoms), $4\sigma_v$ present.	D_{4h}	$E, C_4^1, C_4^2 = C_2, C_4^3, 4\,C_2 \perp C_4, \sigma_h, 4\sigma_v, S_4^1, S_4^3, i$

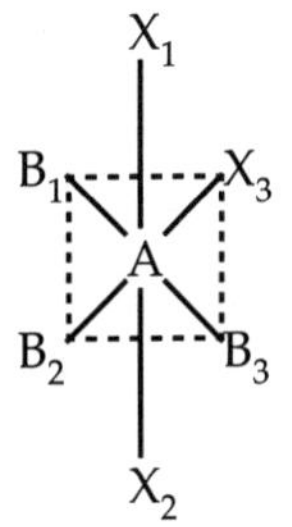	One C_2 present (passes through B_2, A, X_3), $2\sigma_v$ present (one σ_v passes through X_1, X_3, A, B_2, X_2 and the other passes through B_1, B_2, A, B_3, X_3).	C_{2v}	$E, C_2, \sigma_{v1}, \sigma_{v2}$
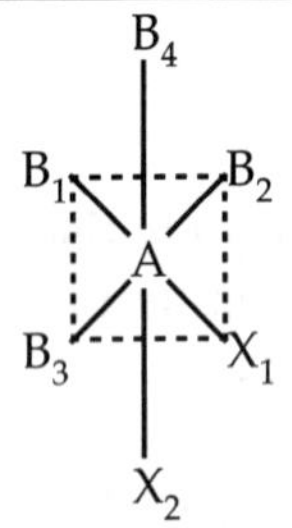	One C_2 present (passes through A and bisects X_1 and X_2 on one side and B_1 and B_4 on the other), $2\sigma_v$ present (one σ_v passes through B_1, B_4, A, X_1, X_2 and the other passes through the above mentioned C_2 axis).	C_{2v}	$E, C_2, \sigma_{v1}, \sigma_{v2}$
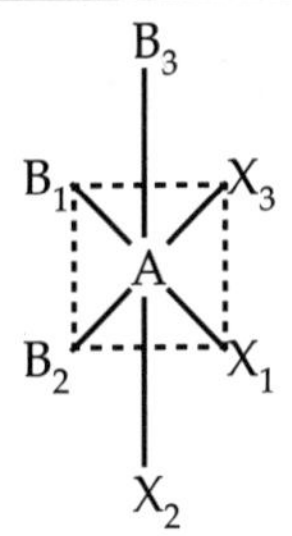	One C_3 axis present (passes through opposite faces), σ_h absent, $3\sigma_v$ present.	C_{3v}	$E, C_3^1, C_3^2, \sigma_{v1}, \sigma_{v2}, \sigma_{v3}$

Table 2.3 Symmetry elements and point groups of molecules

S. No.	Molecule	Search of minimum symmetry elements for identification of point groups	Point group	Complete list of symmetry elements
1.	Acetylene H–C≡C–H	C_∞ axis is present (bond axis), i is present.	$D_{\infty h}$	$E, C_\infty, \infty\perp C_2, \sigma_h, \infty\sigma_v, i$
2.	Carbonyl sulphide S=C=O	C_∞ axis is present (bond axis), no i.	$C_{\infty v}$	$E, C_\infty, \infty\sigma_v$
3.	Water 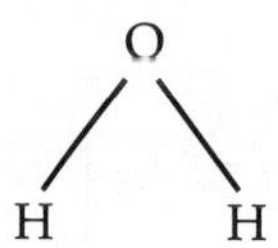	One C_2 axis is present (passes through O and bisects the line joining the two H atoms. No σ_h. $2\sigma_v$ present (molecular plane and the plane passing through O and bisecting the line joining the two H atoms).	C_{2v}	$E, C_{2(Z)}, \sigma_{v1}, \sigma_{v2}$
4.	*Cis*-dichloroethylene 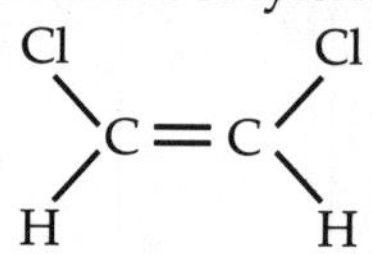	One C_2 axis bisecting the C=C and the lines joining the two chlorine atoms and the two hydrogen atoms, σ_h absent, 2 σ_v present (the molecular plane and the plane bisecting the C=C bond).	C_{2v}	$E, C_2, 2\sigma_v$
5.	*Trans*-dichloroethylene 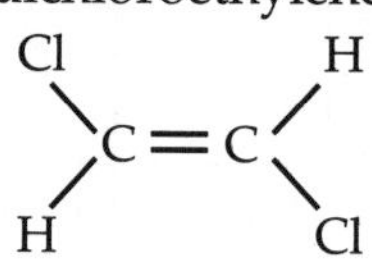	One C_2 axis present (it passes through the centre of the molecule and is ⊥ to the molecular plane) σ_h present (molecular plane).	C_{2h}	E, C_2, σ_h, i
6.	Nitrogen trifluoride 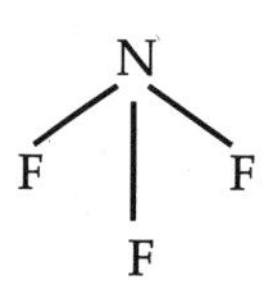	One C_3 axis (passing through nitrogen and the centre of the triangle having fluorines at the corners), σ_h absent, 3 σ_v present (each passing through nitrogen and one of the fluorine atoms and bisecting the line joining the other 2 fluorine atoms).	C_{3v}	$E, C_3, 3\sigma_v$

7.	Methyl chloride 	One C_3 axis (passing through the C and Cl atoms and also through the centre of the triangle formed by joining the three hydrogen atoms), σ_h absent, 3 σ_v present (each passing through C, Cl atoms and one of the H atoms, and bisecting the line joining the other two H atoms).	C_{3v}	$E, C_3^1, C_3^2,$ $3\sigma_v$
8.	Trichloro-methane 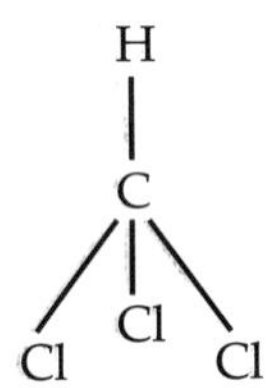	One C_3 axis, σ_h absent, 3 σ_v (as in CH_3Cl).	C_{3v}	E, C_3^1, C_3^2 $3\ \sigma_v$
9.	Dichloro-methane 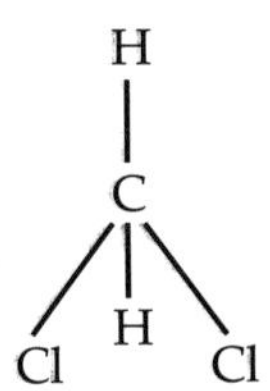	C_2 axis present (passing through C and bisecting the line joining the two Cl atoms on one side and the two H atoms on the other side), 2 σ_v present (one σ_v passing through C and two Cl atoms and bisecting the line joining the two H atoms and the other passing through C and two H atoms and bisecting the line joining the two Cl atoms).	C_{2v}	$E, C_2, 2\ \sigma_v$
10.	Bromochloro-methane 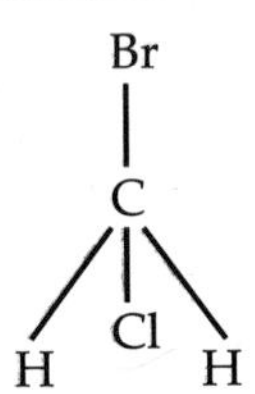	No C_n axis (where n >1), i absent, σ present (the plane passing through C, Cl, Br atoms and bisecting the line joining the two H atoms).	C_s	E, σ

11.	Bromochloro-fluoromethane 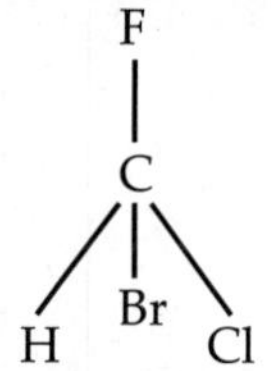	C_n absent (n >1), *i* absent, σ absent.	C_1	*E*
12.	Ethylene 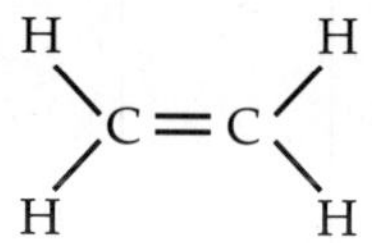	3 C_2 axes present (the principal C_2 axis bisects the C=C bond and is ⊥ to the molecular plane, the second C_2 axis passes through C=C bond and bisects the line joining the adjacent II on both sides, the third C_2 axis bisects the C=C bond and is in the molecular plane), σ_h is the molecular plane.	D_{2h}	E, $3C_2$, σ_h, $2\sigma_v$, i
13.	Chloroethylene 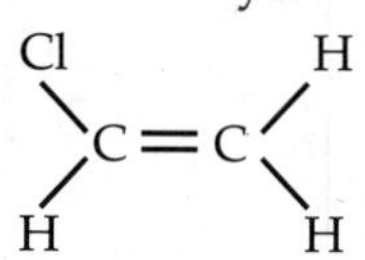	No C_n (where n >1), *i* absent, σ present (molecular plane is the only plane present).	C_s	*E*, σ
14.	Boron trichloride 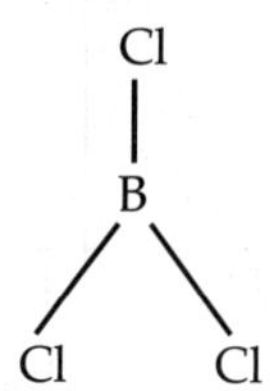	C_3 axis present (passes through B and is ⊥ to the molecular plane), 3 C_2 ⊥ to C_3 present (each C_2 passes through B and one of the Cl atoms and bisects the line joining the other two Cl atoms), σ_h present (molecular plane).	D_{3h}	E, C_3^1, C_3^2, $3\sigma_v$, $3 \perp C_2$, σ_h, S_3^1, S_3^5
15.	$[PtCl_4]$ 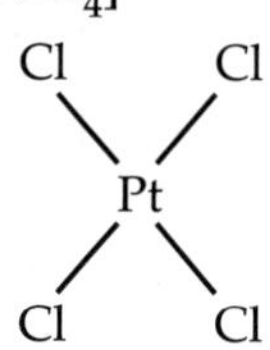	C_4 axis present (passing through Pt and ⊥ to the molecular plane), four ⊥ C_2 axes also present (2 C_2 axes passing through Pt and the diagonally opposite Cl atoms, the other two C_2 axes passing through Pt and bisecting the line joining the adjacent Cl atoms on both	D_{4h}	E, C_4^1, $C_4^2 = C_2$, C_4^3, $4C_2 \perp C_4$, σ_h, $4\sigma_v$, S_4^1, S_4^3, i

		sides), σ_h present (molecular plane), 4 σ_v present (two σ_v passing through Pt and bisecting the line joining the adjacent Cl atoms on both sides, two other σ_v passing through Pt and the diagonally opposite Cl atoms).		
16.	*Cis*-$[PtCl_2Br_2]$ 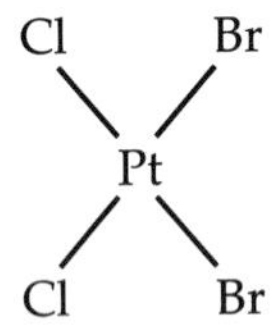	C_2 axis present (passing through Pt and bisecting the line joining the adjacent Cl atoms and adjacent Br atoms), two σ_v present (the molecular plane and the plane passing through Pt and bisecting the line joining the two Cl atoms on one side and two Br atoms on the other).	C_{2v}	$E, C_2, 2\sigma_v$
17.	*Trans*-$[PtCl_2Br_2]$ 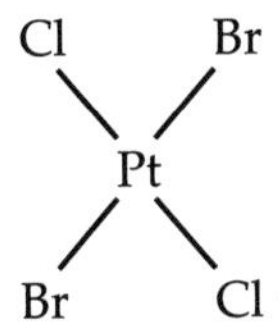	3 C_2 axes present (the principal C_2 axis passes through Pt and is ⊥ to the molecular plane, the second C_2 axis passes through Pt and two diagonally opposite Cl atoms, the third C_2 axis passes through Pt and two diagonally opposite Br atoms), σ_h is the molecular plane, 2 σ_v present.	D_{2h}	$E, C_2, 2C_2 \perp C_2, \sigma_h, 2\sigma_v, i$
18.	Benzene 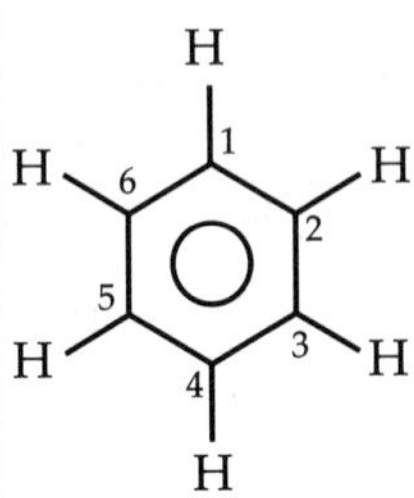	C_6 axis present (passes through the centre of the molecule and is ⊥ to the molecular plane), 6 C_2 ⊥ to C_6 (3 C_2 axes pass through the centre of the molecule and bisect the line joining the adjacent H atoms on both sides, the other three C_2 axes pass through the centre of the molecule and the two opposite H atoms), σ_h is the molecular plane, 6 σ_v are present.	D_{6h}	$E, C_6^1, C_6^2 = C_3, C_6^3 = C_2, C_6^4 = C_3^2, C_6^5, 6\,C_2 \perp C_6, \sigma_h, 6\,\sigma_v, S_6, (C_6^1 \times \sigma_h = S_6), S_6^5, S_3^1, S_3^5, i$

19.	Chlorobenzene 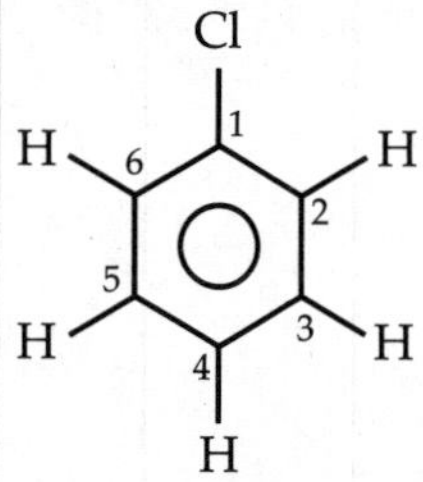	One C_2 axis present (it passes through Cl, C_1 and C_4 atoms), σ_h absent, 2 σ_v present (the molecular plane and the plane passing through the Cl, C_1 and C_4 atoms).	C_{2v}	$E, C_2, 2\ \sigma_v$
20.	Toluene (excluding three methyl hydrogens) 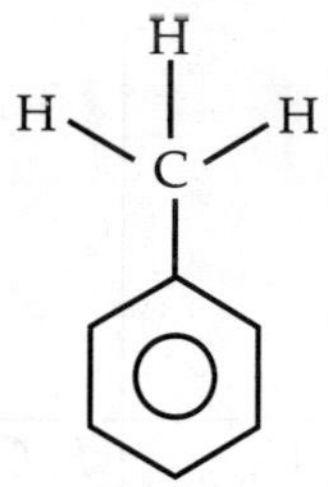	One C_2 axis present (it passes through CH_3, C_1 and C_4 atoms), σ_h absent, 2 σ_v present (the molecular plane and the plane passing through the CH_3 and the diagonally opposite H atom).	C_{2v}	$E, C_2, 2\ \sigma_v$
21.	Toluene (including three methyl hydrogens) 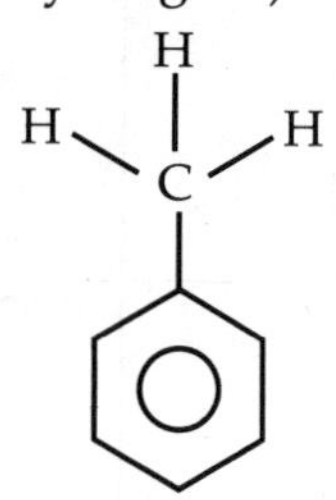	One C_3 axis present (it passes through the phenyl ring and the methyl carbon atom), σ_h absent, $3\sigma_v$ present (each σ_v passes through the phenyl ring, the methyl carbon and one hydrogen atom).	C_{3v}	$E, C_3, 3\sigma_v$
22.	*o*-Dichloro-benzene 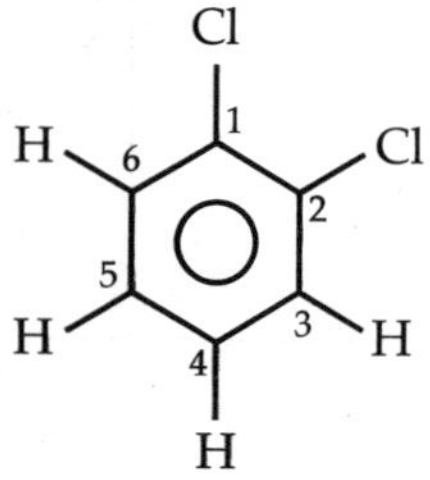	C_2 axis present (it passes through the centre and bisects the line joining C_1, C_2 and C_4, C_5), no σ_h, 2 σ_v present (the molecular plane and the plane passing through the centre and bisecting the line joining C_1, C_2 and C_4, C_5).	C_{2v}	$E, C_2, 2\sigma_v$

23.	*m*-dichloro-benzene 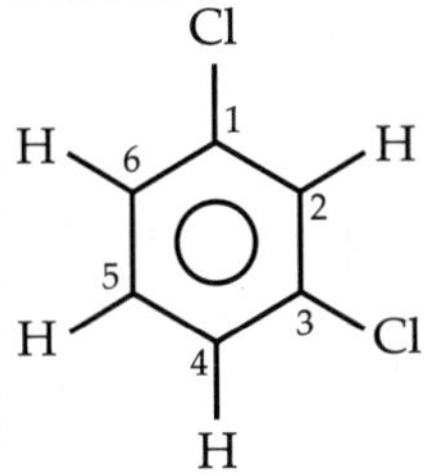	C_2 axis present (it passes through C_2 and C_5 atoms), 2 σ_v present (the molecular plane and the plane passing through C_2 and C_5).	C_{2v}	$E, C_2, 2\sigma_v$
24.	*p*-dichloro-benzene 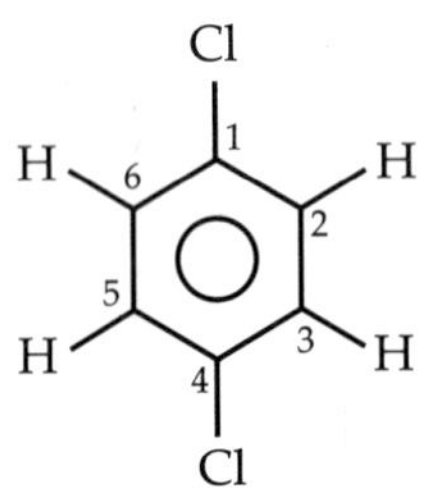	3 C_2 axes present (principal C_2 axis passes through the centre of the molecule and is ⊥ to the molecular plane, the second C_2 axis passes through C_1 and C_4, the third C_2 axis bisects the C_2, C_3 bond on one side and the C_5, C_6 on the other), σ_h is the molecular plane.	D_{2h}	$E, C_2, \sigma_h, 2C_2 \perp C_2, 2\sigma_v, i$
25.	*p*-bromochloro benzene 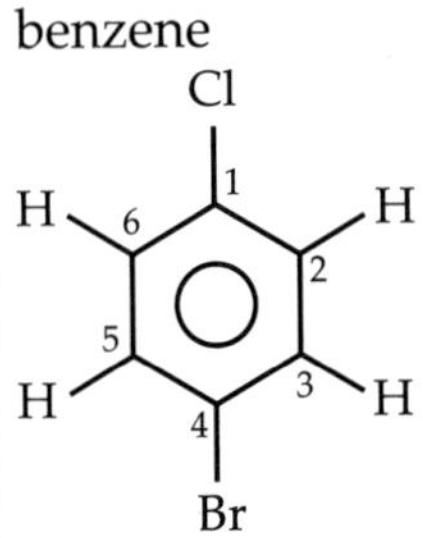	One C_2 axis present (it passes through Cl, C_1, C_4 and the Br atom), no ⊥ C_2, σ_h absent, 2 σ_v present (molecular plane and the plane passing through the Cl, C_1, C_4 and Br atoms).	C_{2v}	$E, C_2, 2\sigma_v$
26.	1,2,3-trichloro-benzene 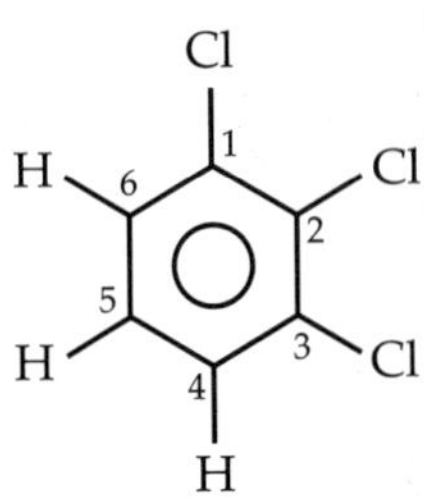	One C_2 present (it passes through Cl, C_2 and C_5 atoms), 2 σ_v present (the molecular plane and the plane passing through Cl, C_2 and C_5).	C_{2v}	$E, C_2, 2\sigma_v$

27.	1,2,4-trichloro-benzene 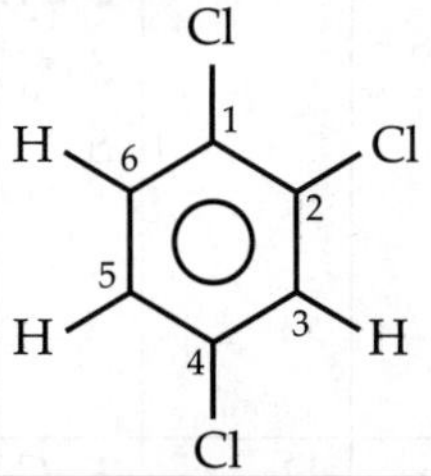	C_n absent, i absent, one σ present (molecular plane).	C_S	E, σ
28.	1,3,5- trichloro-benzene 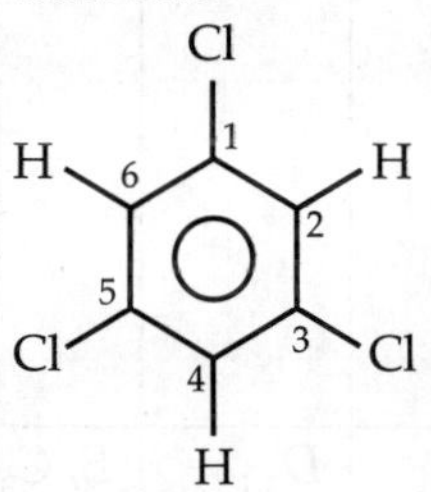	C_3 axis present (it passes through the centre of the molecule and is ⊥ the molecular plane), 3 C_2 ⊥ to C_3 [each C_2 passes through one Cl and the opposite C atom (Cl–C_1–C_4, Cl–C_3–C_5, Cl–C_5–C_2)], σ_h is present (molecular plane), 3 σ_v present.	D_{3h}	$E, C_3^1, 3\sigma_v, C_3^2, 3\perp C_2, \sigma_h, S_3^1, S_3^5$
29.	Naphthalene 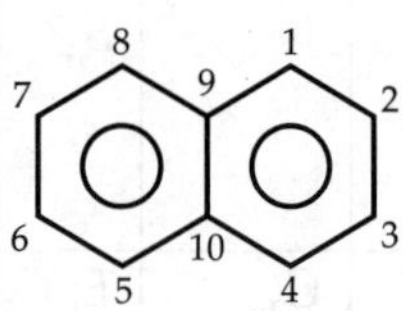	Principal C_2 axis (passes through centre of the molecule and is ⊥ to the molecular plane), 2 C_2 ⊥ to principal C_2 (one C_2 passes through C_9 and C_{10}, the other one passes through the centre of the line joining C_6, C_7, C_2 and C_3), σ_h present (molecular plane).	D_{2h}	$E, 3C_2, \sigma_h, \sigma_{2v}, i$
30.	Pyridine 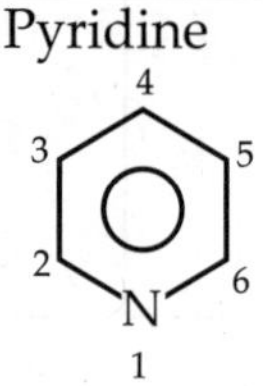	One C_2 present (passes through C_4 and N), 2 σ_v present (the molecular plane and the plane passing through N and C_4).	C_{2v}	$E, C_2, 2\sigma_v$
31.	Eclipsed ethane 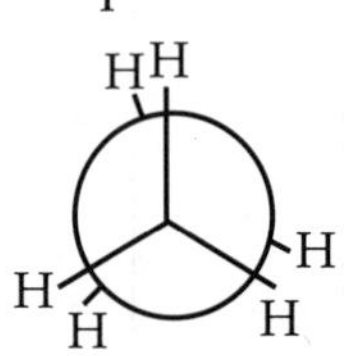	One C_3 present (passes through the C–C bond), 3 C_2 ⊥ to C_3 (every C_2 axis bisects one front and one back H atom on both sides), σ_h is present (passing through the plane bisecting the C–C bond).	D_{3h}	$E, C_3^1, C_3^2, 3C_2 \perp C_3, 3\sigma_v, \sigma_h, S_3^1, S_3^5$

32.	Staggered ethane 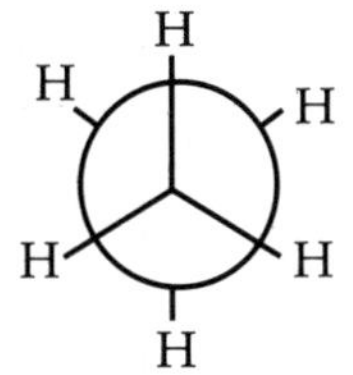	One C_3 present (passes through the C–C bond), 3 C_2 ⊥ to C_3 (every C_2 axis bisects one front and one back H atom on both sides), 3 σ_d (passing through the front –CH and back –CH and bisecting the two C_2 axes).	D_{3d}	E, C_3^1, C_3^2, 3 $C_2 \perp C_3$, 3 σ_d, S_6^1, S_6^5
33.	Skew conformation of ethane 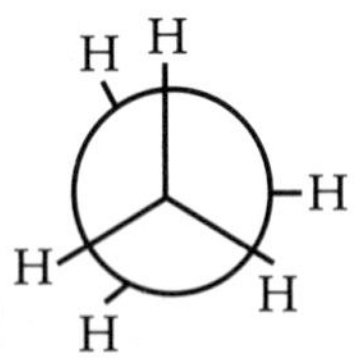	One C_3 present, 3 C_2 ⊥ to the C_3 (bisecting front and back hydrogens on both sides).	D_3	E, C_3^1, C_3^2, 3 $C_2 \perp C_3$
34.	Cyclohexane (chair conformation) 	One C_3 (passing through the centre of the molecule), 3 C_2 ⊥ to C_3 (each C_2 passes through diagonally opposite C–C bonds), 3 σ_d present (passing through diagonally opposite C–H groups).	D_{3d}	E, C_3^1, C_3^2, $3C_2 \perp C_3$, 3 σ_d, S_6^1, S_6^5
35.	Cyclohexane (boat) 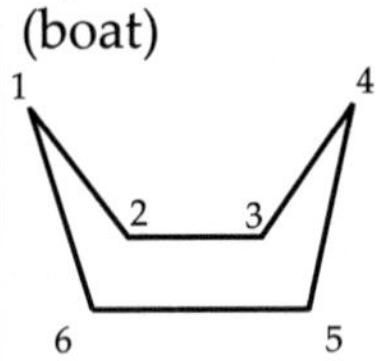	One C_2 (passing through the centre of the molecule), 2 σ_v (one plane bisects the C2–C3 bond on one side and the C5–C6 on the other, the second plane passes through the center of the molecule, C1 and C4 and bisects the molecule).	C_{2v}	E, C_2, 2 σ_v
36.	Norbornane 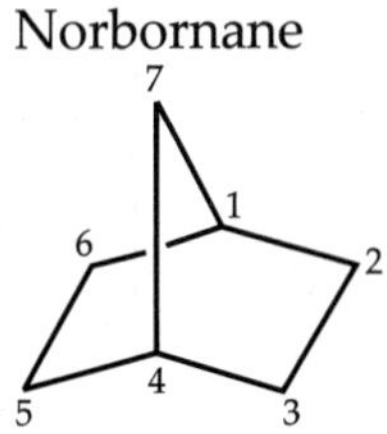	One C_2 present (passing through the bridge carbon atom and the centre of the molecule), 2 σ_v present (one plane passing through C1, C7 and C4, and the other passing through C7 and bisecting the C2–C3 bond on one side and the C5–C6 bond on the other).	C_{2v}	E, C_2, 2 σ_v

37.	ML_6 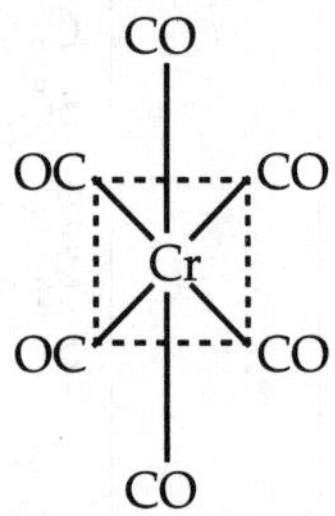	3 C_4 axes present (each one passing through M and 2 diagonally opposite ligands), $4C_3$ axes present (each passing through M and the opposite faces of the octahedron), 6 C_2 present (each passing through M and the opposite edges of the octahedron), 3 σ_h, 6 σ_v passing through opposite faces.	O_h	$E, 3C_4^1, 3C_4^2, 3C_4^3, 4C_3^1, 4C_3^2, 6C_2, 3\sigma_h, 6\sigma_v$, i, $3S_4^1, 3S_4^5, 4S_6^1, 4S_6^5$
38.	*Cis*-$[ML_4X_2]$ 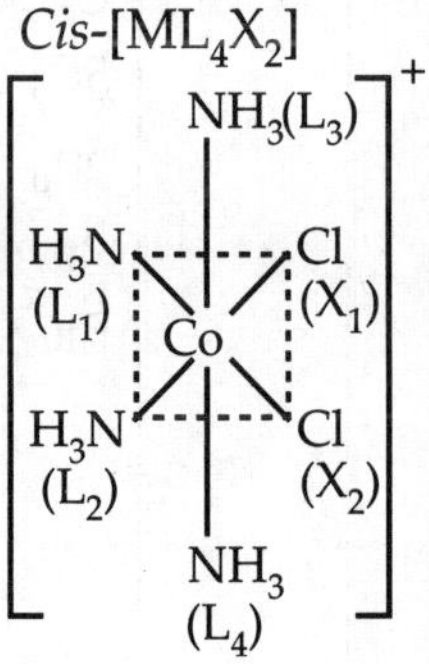	One C_2 axis present (passing through M and bisecting the lines joining L_1 and L_2 on one side and X_1 and X_2 on the other), 2 σ_v present (one passing through L_3–M–L_4 and bisecting the line L_1–L_2 on one side and X_1–X_2 on the other side, and another passing through L_1–L_2–M–X_1–X_2).	C_{2v}	$E, C_2, 2\sigma_v$
39.	*Trans*-$[ML_4X_2]$ 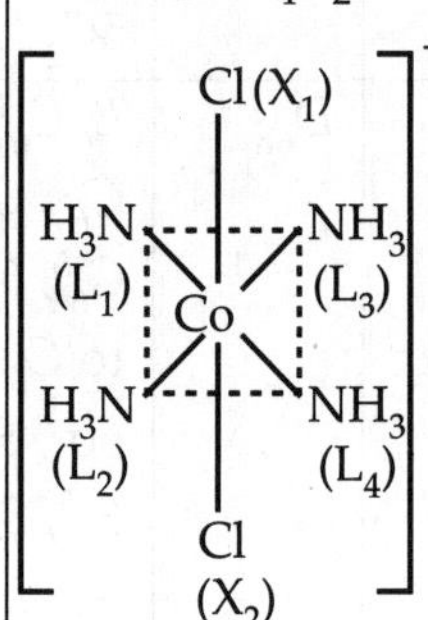	One C_4 present (passing through X_1–M–X_2), 4 $C_2 \perp$ to C_4 present (two C_2 axes passing through M and bisecting the line joining adjacent L atoms on both sides and two C_2 axes passing through M and diagonally opposite L atoms), σ_h present (square plane passing through M and four L), 4 σ_v present (planes replacing C_2 axes).	D_{4h}	$E, C_4^1, C_4^2=C_2, C_4^3, 4C_2 \perp$ to $C_4, \sigma_h, 4\sigma_v, S_4^1, S_4^3, i$

40.	Ferrocene (eclipsed) 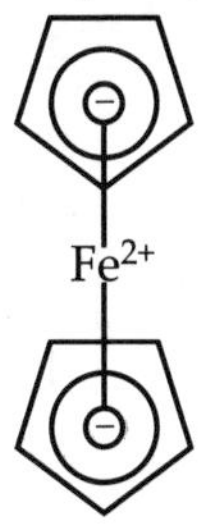	One C_5 present (passing through Fe and the centre of the two pentagons), 5 $C_2 \perp$ to C_5 present (each C_2 passes through Fe and bisects the front and back carbon atoms), σ_h present (passes through Fe and bisects the two rings), 5 σ_v present (planes replacing C_2 axes).	D_{5h}	$E, C_5^1, C_5^2, C_5^3, C_5^4, 5C_2 \perp C_5, \sigma_h, 5\sigma_v, S_5^1, S_5^3, S_5^7, S_5^9$
41.	Ferrocene (staggered) 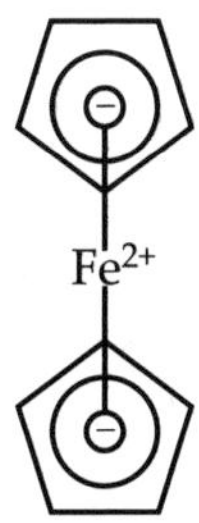	One C_5 present (passing through Fe and the centre of the two pentagons), 5 $C_2 \perp$ to C_5 present (each C_2 passes through opposite C atoms and Fe), σ_h absent, 5 σ_d present	D_{5d}	$E, C_5^1, C_5^2, C_5^3, C_5^4, 5C_2 \perp C_5, 5\sigma_d, S_{10}^1, S_{10}^3, S_{10}^5, S_{10}^7, S_{10}^9$
42.	Ferrocene (neither eclipsed nor staggered)	One C_5 present, 5 $C_2 \perp$ to C_5 present (as before), σ_h absent, σ_v absent.	D_5	$E, C_5^1, C_5^2, C_5^3, C_5^4, 5C_2 \perp C_5$
43.	Diphenyl-chromium (eclipsed)	One C_6 present (passing through Cr and the centres of the two hexagons), 6 $C_2 \perp$ to C_6 (each passing through Cr, the front and back C atoms on both sides), σ_h present (passing through Cr and bisecting the ring), 6 σ_v present (planes replacing C_2 axes).	D_{6h}	$E, C_6^1, C_6^2, C_6^3, C_6^4, C_6^5, 6C_2 \perp$ to $C_6, \sigma_h, 6\sigma_v, S_6^1, S_6^5, S_3^1, S_3^5, i$
44.	Diphenyl-chromium (staggered)	One C_6 present (passes through Cr and the centres of the two hexagons), 6 $C_2 \perp$ to C_6 present (passing through opposite C atoms and Cr), σ_h absent, 6 σ_d present.	D_{6d}	$E, C_6^1, C_6^2, C_6^3, C_6^4, C_6^5, 6C_2 \perp$ to $C_6, 6\sigma_d, S_{12}^1, S_{12}^3, S_{12}^5, S_{12}^7, S_{12}^9, S_{12}^{11}$

45.	Biphenyl (planar) 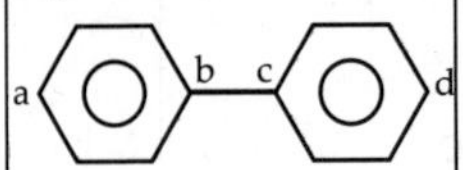	C_2 present (passing through centre of the molecule and ⊥ to the molecular plane—principal axis), 2 C_2 ⊥ to principal C_2 [one passing through the centre of the molecule in the molecular plane and other through four carbon atoms (a, b, c, d)], σ_h present (molecular plane), 2 σ_v present.	D_{2h}	$E, 2\sigma_v, 3C_2, \sigma_h, i$
46.	Glyoxal 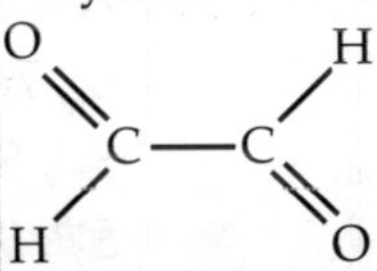	C_2 axis bisects the C–C bond and is perpendicular to the molecular plane, σ_h present.	C_{2h}	E, C_2, σ_h, i
47.	Borazine 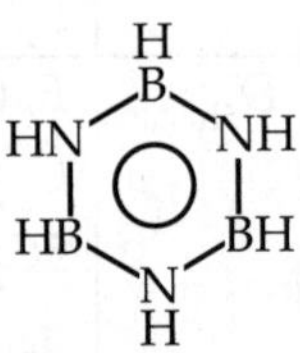	C_3 axis passes through center of the molecule and is perpendicular to the molecular plane, three C_2 axes pass through opposite B–H and N–H groups, σ_h present (molecular plane).	D_{3h}	$E, C_3^1, C_3^2, 3C_2 \perp C_3, 3\sigma_v, \sigma_h, S_3^1, S_3^5$
48.	Unsymmetrical biphenyl 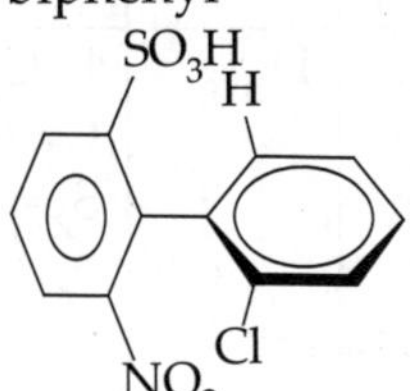	Only identity element	C_1	E
49.	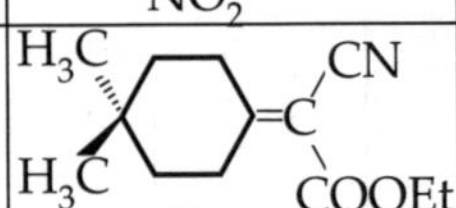	Only identity element	C_1	E
50.	Spiran 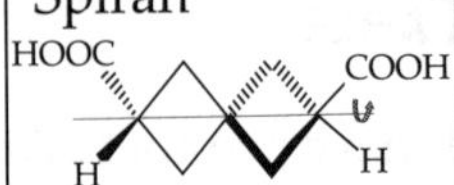	Only C_2 axis as shown.	C_2	E, C_2

51.	2,6 dichloro quinone 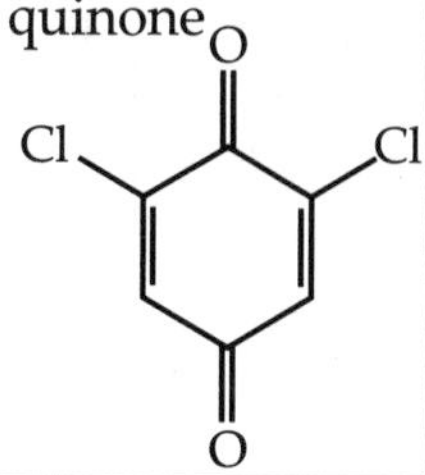	C_2 axis passes through opposite carbonyl groups, two σ_v present (one is molecular plane, another passes through the opposite carbonyl groups).	C_{2v}	$E, C_2, 2\sigma_v$
52.	Tropylium ion 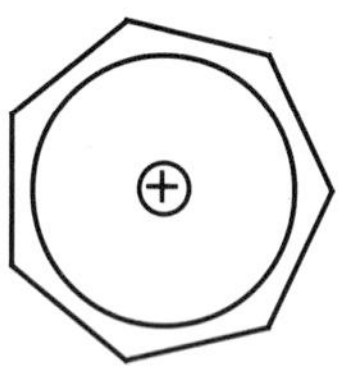	C_7 axis passes through the centre and is perpendicular to the molecular plane, seven C_2 axes pass through CH and the mid point of the opposite C– C bond, σ_h is the molecular plane, seven σ_v present.	D_{7h}	$E, C_7^1, C_7^2, C_7^3, C_7^4, C_7^5, C_7^6, 7C_2 \perp C_7, \sigma_h, 7\sigma_v, S_7^1, S_7^3, S_7^5, S_7^7, S_7^9, S_7^{11}, S_7^{13}$
53.	Cyclopenta-dienyl cation 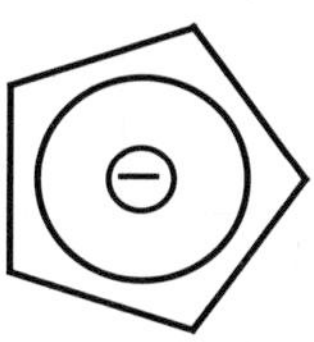	C_5 axis passes through the centre and is perpendicular to the molecular plane, five C_2 axes pass through CH and the mid point of opposite C–C bond, σ_h is the molecular plane, five σ_v present.	D_{5h}	$E, C_5^1, C_5^2, C_5^3, C_5^4, 5C_2 \perp C_5, \sigma_h, 5\sigma_v, S_5^1, S_5^3, S_5^7, S_5^9$
54.	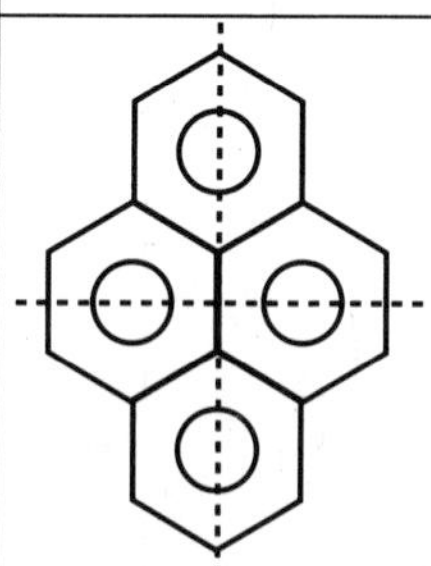	Principal C_2 axis passes through the center of the molecule and is perpendicular to the molecular plane, two C_2 axes perpendicular to the principal C_2 axis as shown, σ_h is the molecular plane.	D_{2h}	$E, C_2, 2 \perp C_2, \sigma_h, \sigma_v, S_2 = i$

2.4 GROUP MULTIPLICATION TABLE

A group multiplication table is a table which gives the binary product of any two members of the group and the square of every member of a group. The first row and first column of the group multiplication table list all symmetry operations of a particular point group one after another in the same order. The other rows and columns of the table indicate the binary product of the element in a row followed by the element in a column. Each point group has a group multiplication table that shows the interrelationships between the elements. This can be derived by determining the product of the symmetry elements as discussed in the last chapter. The group multiplication tables of point groups C_{2V} and C_{3V} are shown in Tables 2.4 and 2.5.

Table 2.4 Group multiplication table for C_{2V}

C_{2V}	E	C_2	σ_{V1}	σ_{V2}
E	E	C_2	σ_{V1}	σ_{V2}
C_2	C_2	E	σ_{V2}	σ_{V1}
σ_{V1}	σ_{V1}	σ_{V2}	E	C_2
σ_{V2}	σ_{V2}	σ_{V1}	C_2	E

Table 2.5 Group multiplication table for C_{3v}

C_{3V}	E	C_3^1	C_3^2	σ_{V1}	σ_{V2}	σ_{V3}
E	E	C_3^1	C_3^2	σ_{V1}	σ_{V2}	σ_{V3}
C_3^1	C_3^1	C_3^2	E	σ_{V2}	σ_{V3}	σ_{V1}
C_3^2	C_3^2	E	C_3^1	σ_{V3}	σ_{V1}	σ_{V2}
σ_{V1}	σ_{V1}	σ_{V3}	σ_{V2}	E	C_3^2	C_3^1
σ_{V2}	σ_{V2}	σ_{V1}	σ_{V3}	C_3^1	E	C_3^2
σ_{V3}	σ_{V3}	σ_{V2}	σ_{V1}	C_3^2	C_3^1	E

2.5 PROPERTIES OF GROUPS

2.5.1 Order of a Group

The total number of symmetry elements (non-redundant) present in a point group is called the *order of a group*. It is designated as 'h'. For example, the total number of symmetry elements present in a C_{3V} point group is 6 (E, C_3^1, C_3^2, σ_{V1}, σ_{V2}, σ_{v3}) and thus the order, h = 6.

2.5.2 Conjugate Elements and Classes

If a symmetry element A can be transformed to another symmetry element B by means of a similarity transformation done with a third symmetry element C, all belonging to the same point group, then the symmetry elements A and B are referred to as *conjugate elements.* A complete set of elements that are conjugate to one another is called the *class of the group*. The number of classes present in a point group is indicated by 'k'.

The similarity transformation is a mathematical operation involving multiplication by an element on one side and by the inverse of the same element on the other side. For example, when a similarity transformation is done on symmetry element A by a symmetry element C, it can be represented as $C^{-1}AC$. If this ternary product leads to a different symmetry element B, then the elements A and B are referred to as *conjugate elements,* which can be mathematically represented as $C^{-1}AC = B$.

Example 2.1

The symmetry elements present in a C_{3V} point group are E, C_3^1, C_3^2, σ_{V1}, σ_{V2}, σ_{V3}. C_3^1 can be transformed to C_3^2 through a similarity transformation done with σ_{V1}, which is a symmetry element of the same point group. This is illustrated by the NH_3 molecule (Fig. 2.1).

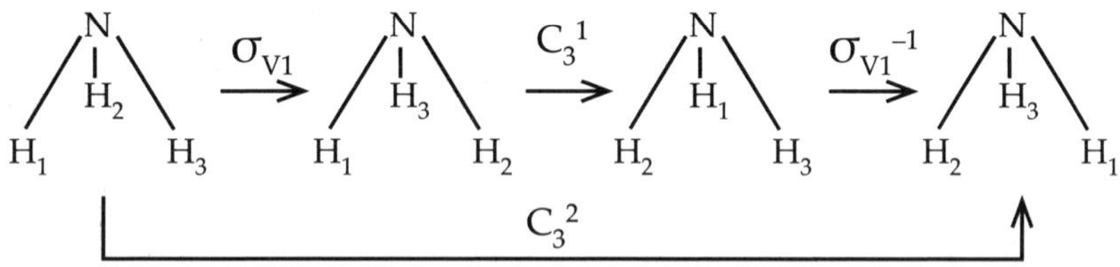

Fig. 2.1 Similarity transformation on NH_3

σ_{V1} is the plane passing through nitrogen and the hydrogen on the left, H_1. σ_{V2} is the plane passing through nitrogen and hydrogen on the right, H_3. σ_{V3} is the plane passing through nitrogen and central hydrogen, H_2.

Since the ternary product $\sigma_V^{-1}C_3^1\sigma_{v1} = C_3^2$, C_3^1 and C_3^2 are conjugate elements. Similarly, it can be shown that σ_{V1} and σ_{V2} of the C_{3V} point group are conjugate elements (Fig. 2.2).

That is, $C_3^{-1}\ \sigma_{V1}\ C_3^1 = \sigma_{V2}$

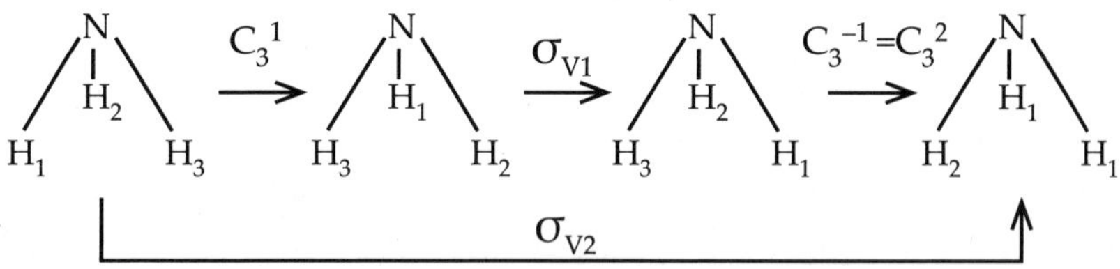

Fig. 2.2 Similarity transformation on NH_3

It can also be shown that (i) σ_{V2} and σ_{V3} and (ii) σ_{V1} and σ_{V3} are conjugate elements.

(i) σ_{V2} and σ_{V3} (Fig. 2.3)

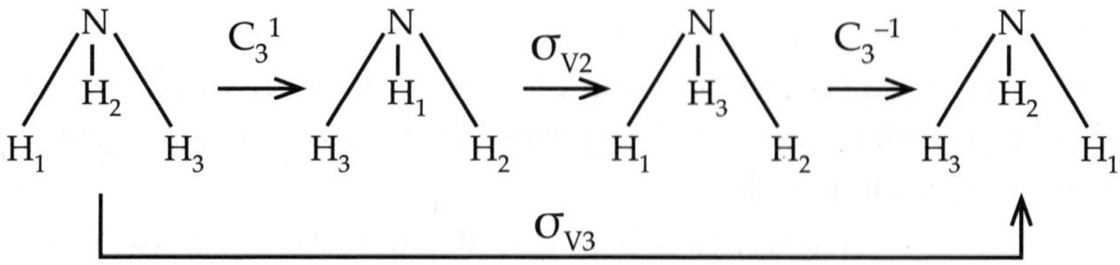

Fig. 2.3 Similarity transformation on NH_3

(ii) σ_{V2} and σ_{V3} (Fig. 2.4)

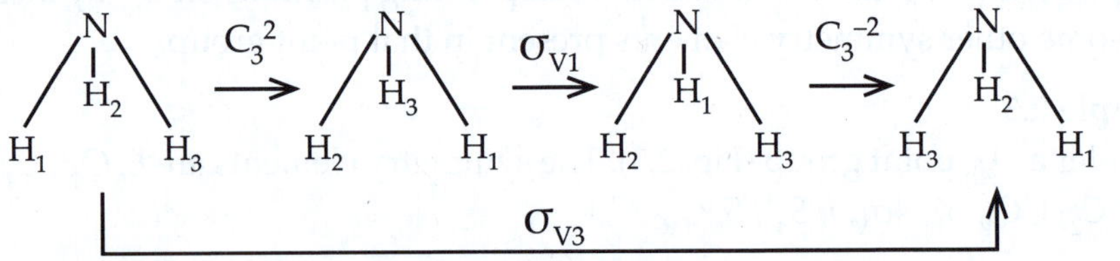

Fig. 2.4 Similarity transformation on NH_3

Thus, the number of classes (k) in the C_{3V} group is 3. They are E, $2C_3$ and $3\sigma_V$.

Note: The element C with respect to which the similarity transformation is done to identify the conjugate element should not commute with either element A or B. If they commute, similarity transformation will lead to the same element instead of the conjugate element.

That is, $C^{-1}A\,C = C^{-1}C\,A = EA = A$

Example 2.2

Consider a C_3 point group. The symmetry elements of this group are E, C_3^1, C_3^2. The order of the group is 3 (h=3). C_3^1 and C_3^2 commute with the identity element E and hence C_3^1 cannot be transformed into C_3^2 by a similarity transformation. There are no other elements in the group to check for conjugate elements. Thus, they belong to different classes. Therefore, in the C_3 point group, h = 3 and k = 3.

The above method which uses the equation $C^{-1}AC = B$ to identify the conjugate elements and classes of a point group is cumbersome. For example, to identify the number of classes present in the D_{3h} point group which has 12 symmetry elements, similarity transformation operations have to be worked out 144 times. Hence, there should be a simpler way to identify the number of classes in a point group. This can be achieved by using the following method based on similarity transformation equation.

2.5.3 Simple Method to Identify the Classes Present in a Point Group

When a few simple rules are kept in mind and applied, the classes present in a point group can be easily identified.

(i) E, σ_h and i commute with all symmetry elements and as a result they form separate classes in every point group.

(ii) The principal axis C_n and its inverse C_n^{-1} belong to the same class, if a σ_V is present in the point group or a C_2 axis perpendicular to C_n is present.

(iii) The symmetry elements S_n and S_n^{-1} are conjugate symmetry elements and can be clubbed into the same class if a σ_V is present in the molecule or a C_2 axis perpendicular to S_n is present.

That is, $\sigma_V^{-1}\,S_n\,\sigma_V = S_n^{-1}$

4. Other axes C_m and $C_{m'}$ will belong to the same class if the atoms or groups present on C_m can be moved to corresponding positions on $C_{m'}$ by means of some other symmetry elements present in that point group.

Example 2.3

Consider a D_{4h} point group (Fig. 2.5). The symmetry elements are E, C_4^1, $C_4^2 = C_2$, C_4^3, $4\,C_2 \perp C_4$, σ_h, $4\sigma_V$, i, S_4^1, S_4^3.

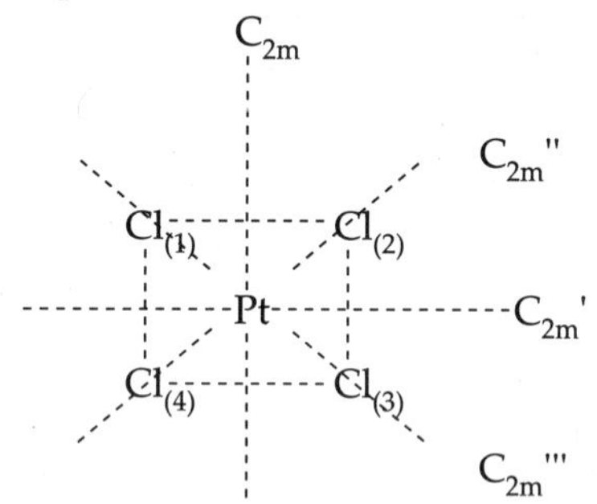

Fig. 2.5

C_{2m}'' and C_{2m}''' belong to one class since the atoms on the C_{2m}'' axis can be moved to the C_{2m}''' by the C_4^1 operation. C_{2m} and C_{2m}' belong to the same class since both contain only the platinum atom. C_{2m} and C_{2m}'' belong to different classes because C_{2m} has only Pt and C_{2m}'' has Pt and two chlorine atoms. So, the atoms present on C_{2m}'' cannot be moved to the C_{2m} axis. They are not conjugate symmetry elements.

5. Two symmetry planes σ_{V1} and σ_{V2} will belong to the same class, if the atoms or groups present in σ_{V1} can be moved to σ_{V2} by some symmetry element present in the point group (Fig. 2.6).

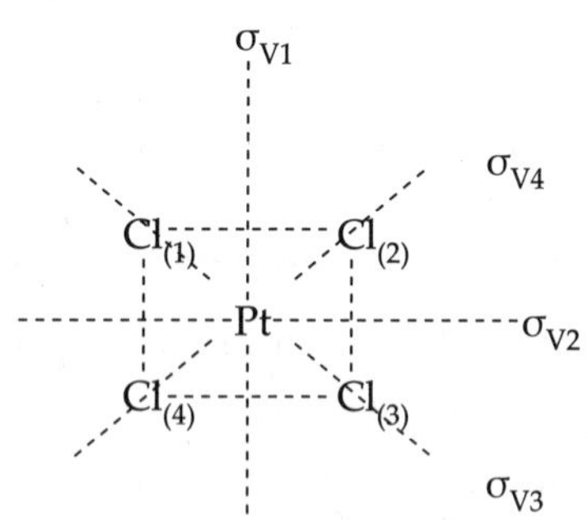

Fig. 2.6

These two elements belong to the same class as both contain only Pt. The atoms on σ_{V3} can be moved to σ_{V4} by doing a C_4^1 operation and hence σ_{V3} and σ_{V4} are conjugate elements and belong to the same class. Since the planes σ_{V1} and σ_{V2} contain only the Pt atom, and the planes σ_{V3} and σ_{V4} contain the Pt and two chlorine atoms, transformation of σ_{V1} to σ_{V4} is not possible. Hence, they are not conjugate elements.

Thus, by applying the above rules, it is possible to determine the number of classes present in various point groups.

Example 2.4

Determine the number of classes present in some important point groups.

(i) C_{2V} — $E, C_2, \sigma_{v1}, \sigma_{v2}$
h = 4, k = 4 since all the elements commute with each other.

(ii) C_{2h} — E, C_2, σ_h, i
h = 4, k = 4.

(iii) C_{3h} — $E, C_3^1, C_3^2, \sigma_h, S_3^1, S_3^5$
h = 6, k = 6. E and σ_h form separate classes. C_3^1 and C_3^2 belong to different classes since σ_V is absent.

(iv) C_{4v} — $E, C_4^1, C_4^2 = C_2, C_4^3, \sigma_{V1}, \sigma_{V2}, \sigma_{V3}, \sigma_{V4}$
h = 8, k = 5 [The classes are, $E, (C_4^1, C_4^3), C_4^2 = C_2, 2\,\sigma_V, 2\,\sigma_V$].

(v) D_{2h} — $E, C_2, 2\,C_2 \perp C_2, \sigma_h, 2\,\sigma_v$, i
h = 8, k = 8.

(vi) D_{3h} — $E, C_3^1, C_3^2, 3\,C_2 \perp C_3, \sigma_h, 3\,\sigma_v, S_3^1, S_3^5$
h = 12, k = 6 [The classes are, $E, (C_3^1, C_3^2), 3\,C_2 \perp C_3, \sigma_h, 3\sigma_v, (S_3^1, S_3^5)$].

(vii) D_{6h} — $E, C_6^1, C_6^2 = C_3, C_6^3 = C_2, C_6^4 = C_3^2, C_6^5, 6\,C_2 \perp C_6, \sigma_h, 6\,\sigma_v, S_6^1, S_6^3, S_6^5, S_3^1, S_3^5$, i
h =12,k = 24 [The classes are, $E, \sigma_h, (C_6^1$ and $C_6^5), 3\,C_2, 3\,C_2, 3\,\sigma_v, 3\,\sigma_v, 2\,S_6, 2\,S_3, (C_3^1$ and $C_3^2), (C_6^2$ and $C_6^4), C_6^3 = C_2, i$].

(viii) T_d — $E, 4\,C_3^1, 4\,C_3^2, 3\,C_2, 6\,\sigma_{v1}, 3\,S_4^1, 3\,S_4^3$
h = 24, k = 5 [The classes are, $E, 8\,C_3, 6\,\sigma_v, 6\,S_4, 3\,C_2$].

(ix) O_h — $E, 3\,C_4^1, 3\,C_4^2, 3\,C_4^3, 4\,C_4^1, 4\,C_4^2, 6\,C_2, 3\,\sigma_h, 6\,\sigma_v$, i, $4\,S_6^1, 4\,S_6^5, 3\,S_4^1, 3\,S_4^3$
h = 48, k =10 [The classes are, $E, 3\,\sigma_h$, i, $6\,C_4, 3\,C_4^2, 8\,C_3, 6\,C_2, 6\,\sigma_v, 8\,S_6, 6\,S_4$].

2.5.4 Subgroups

If a lesser number of symmetry elements of a group satisfy all the four mathematical rules for the formation of a group, they form a subgroup of the main group. A subgroup must satisfy the following conditions.

(i) Order of a subgroup should be less than that of the main group.

(ii) The ratio of the order of main group to the order of subgroup should be an integer.

That is, h = n × p; where 'h' is the order of the main group, 'p' is the order of the sub group and 'n' is a whole number.

Example 2.5

Consider the D_{2h} point group. The symmetry elements of this group are $E, C_{2(z)}, C_{2(x)}, C_{2(y)}, \sigma_h, \sigma_{v1}, \sigma_{v2}$, i. Here h = 8 and k = 8. The subgroups of D_{2h} are given in Table 2.6.

Table 2.6 Subgroups of D_{2h}

Symmetry elements	Subgroup	Order of subgroup
E	C_1	p = 1
E, C_2	C_2	p = 2
E, σ	C_s	p = 2
E, i	C_i	p = 2
$E, C_{2(Z)}, \sigma_h, i$	C_{2h}	p = 4
$E, C_{2(Z)}, \sigma_{V1}, \sigma_{V2}$	C_{2v}	p – 4
$E, C_{2(x)}, C_{2(y)}, C_{2(Z)}$	D_2	p = 4

Example 2.6

Consider the D_{3h} point group. For this group, h =12 and k = 6. The symmetry elements are $E, C_3^1, C_3^2, C_{2(z)}, C_{2(x)}, C_{2(y)}, \sigma_h, \sigma_{v1}, \sigma_{v2}, \sigma_{v3}, S_3^1, S_3^5$. The subgroups of the D_{3h} group are shown in Table 2.7.

Table 2.7 Subgroups of D_{3h}

Symmetry elements	Subgroup	Order of subgroup
E	C_1	p = 1
E, C_2	C_2	p = 2
E, σ_h	C_s	p = 2
E, C_3^1, C_3^2	C_3	p = 3
$E, C_3^1, C_3^2, 3\sigma_v$	C_{3v}	p = 6
$E, C_3^1, C_3^2, \sigma_h, S_3^1, S_3^5$	C_{3h}	p = 6
$E, C_3^1, C_3^2, 3C_2 \perp C_3$	D_3	p = 6

Example 2.7

Consider the D_{4h} point group. For this group, h =16 and k = 10. The symmetry elements are $E, C_4^1, C_4^2, C_4^3, 4C_2 \perp C_4, \sigma_h, 4\sigma_v, S_4^1, S_4^3, i$. The sub groups of the D_{4h} point group are as shown in Table 2.8.

Table 2.8 Subgroups of D_{4h}

Symmetry elements	Subgroup	Order of subgroup
E	C_1	p = 1
E, C_2	C_2	p = 2
E, σ_h	C_s	p = 2
E, i	C_i	p = 2
E, C_4^1, C_4^2, C_4^3	C_4	p = 4
$E, C_4^1, C_4^2, C_4^3, 4\sigma_v$	C_{4v}	p = 8

$E, C_4^1, C_4^2, C_4^3, \sigma_h, S_4^1, S_4^3, i$	C_{4h}	p = 8
$E, C_2, 2\sigma_v$	C_{2v}	p = 4
E, C_2, σ_h, i	C_{2h}	p = 4
$E, C_2, 2C_2 \perp C_2$	D_2	p = 4
$E, C_4^1, C_4^2, C_4^3, 4C_2 \perp C_4$	D_4	p = 8
$E, C_{2(z)}, 2C_2 \perp C_{2(z)}, \sigma_h, 2\sigma_v, i$	D_{2h}	p = 8

2.5.5 Isomorphic Groups

Two groups are said to be *isomorphic* if they satisfy the following conditions.

(i) The order and number of classes of the two groups must be the same.

(ii) They must produce similar group multiplication tables. That is, they must have one to one correspondence with each other. For example, consider the point groups C_s and C_2. Their group multiplication tables indicate 1:1 correspondence as shown below.

C_s	E	σ
E	E	σ
σ	σ	E

C_2	E	C_2
E	E	C_2
C_2	C_2	E

2.5.6 Abelian Group

An *abelian group* is a group where all the symmetry elements commute with each other. Diagonal elements are identity elements in such a group. For example, the C_{2v} group is an abelian group. Here, all the symmetry elements commute with each other and it is also seen that the diagonal elements are identity elements as indicated by the group multiplication table.

C_{2v}	E	C_2	σ_{v1}	σ_{v2}
E	E	C_2	σ_{v1}	σ_{v2}
C_2	C_2	E	σ_{v2}	σ_{v1}
σ_{v1}	σ_{v1}	σ_{v2}	E	C_2
σ_{v2}	σ_{v2}	σ_{v1}	C_2	E

2.6 APPLICATION OF POINT GROUP ASSIGNMENT

2.6.1 Prediction of Optical Activity

Many substances can rotate a plane of polarised light either towards the right or towards the left. This property is termed *optical activity*. The condition to be satisfied for a molecule to exhibit optical activity is that it should not be superimposable on its mirror image (chirality). This condition is definitely related to the symmetry of the molecule. A molecule which contains an improper axis of symmetry of order 'n' (that is, the S_n axis) is always superimposable on its mirror image as S_n is rotation followed by reflection. Hence the most general rule is that *a molecule having an S_n axis is optically inactive*. Recall here that, $S_1 = \sigma \times C_1 = \sigma$ and $S_2 = \sigma_{xy} \times C_{2(z)} = i$.

Thus, if a molecule belongs to the C_n or the D_n point groups, it can exhibit optical activity as it does not contain any S_n operation.

Optically active molecules can be generally classified as asymmetric or dissymmetric molecules. Asymmetric molecules are devoid of symmetry planes, centre of symmetry and symmetry axes. Such molecules belong to the C_1 point group. Thus molecules with *chiral* carbon(s) are asymmetric molecules and they belong to the C_1 point group and exhibit optical activity. For example, lactic acid and tartaric acid. Dissymmetric molecules exhibit optical activity due to the presence of a chiral axis or chiral planes. Such molecules can have symmetry axes and thus they belong to the C_n or the D_n point groups with n > 1. For example, unsymmetrically substituted allenes and biphenyls.

2.6.2 Point Groups and Dipole Moment

Molecules belonging to point groups having an inversion centre (i) cannot have a dipole moment since the dipole vector cannot lie on the centre. Similarly, molecules with more than one non-coincident axis of rotation cannot have a dipole moment since the dipole vector cannot coincide with two distinct axes. Therefore, molecules belonging to C_i, C_{nh}, I_h and O_h (possessing i) and T_d, D_n, D_{nh}, D_{nd} point groups (possessing more than one non-coincident axis) will have zero dipole moment. Molecules belonging to C_1, C_s, C_n and C_{nv} point groups will have measurable dipole moments.

2.6.3 Symmetry Number

Symmetry number of a molecule indicates the number of equivalent configurations that can be obtained by rotating the molecule around one or more axes. For example, an AB_3 planar molecule has one C_3 axis and three C_2 axes. When the molecule is rotated about the C_3 axis, three equivalent configurations can be obtained. That is, C_3^1, C_3^2, $C_3^3 = E$. When C_2 operations are carried out around each C_2 axis, two equivalent configurations can be obtained—C_2^1 and $C_2^2 = E$, but $C_2^2 = E$ cannot be counted again as it has already been included as $C_3^3 = E$. Thus, rotation about three C_2 axes develops only three equivalent configurations. Therefore, the AB_3 planar molecule has 3 + 3 = 6 equivalent configurations. Symmetry numbers are related to point groups as given in Table 2.9.

Table 2.9 Relationship between point group and symmetry number

Point groups	Symmetry number
$C_{\infty}, C_1, C_s, C_i$	1
C_n, C_{nh}, C_{nv}	n (n is the order of axis)
D_n, D_{nh}, D_{nd}	$2n$
$D_{\infty h}$	2
T_d	12
O_h	24

3

Matrix Representation of Symmetry Operations

It is possible to represent symmetry operations using matrices. Therefore, one must have a basic knowledge about matrices and their characteristics.

3.1 Matrices

A matrix is an array of numbers arranged in rows and columns. A square matrix is a matrix that has an equal number of rows and columns. Consider A which is a 3 × 3 matrix.

$$A = \begin{pmatrix} a_{11} & a_{12} & a_{13} \\ a_{21} & a_{22} & a_{23} \\ a_{31} & a_{32} & a_{33} \end{pmatrix}$$

Here, a_{ij} represents a matrix element in the ith row and the jth column. A is a notation used to represent the matrix A. For each square matrix A with entries that are real numbers, the determinant of A is denoted by ◎A◎ and is given by,

$$\det A = |A| = \Sigma \pm a_{1j_1} a_{2j_2} \ldots a_{nj_n}$$

where, the sum is taken over all possible permutations on n elements and the sign is positive if the number of permutations is even and negative if the number of permutations is odd. For example,

The determinant of the 2 × 2 matrix $A = \begin{pmatrix} a_1 & b_1 \\ a_2 & b_2 \end{pmatrix}$ is defined as $|A| = a_1b_2 - a_2b_1$

and the determinant of a 3 × 3 matrix

$$B = \begin{pmatrix} a_1 & b_1 & c_1 \\ a_2 & b_2 & c_2 \\ a_3 & b_3 & c_3 \end{pmatrix} \text{ is defined as}$$

$$|B| = a_1 \begin{vmatrix} b_2 & c_2 \\ b_3 & c_3 \end{vmatrix} - b_1 \begin{vmatrix} a_2 & c_2 \\ a_3 & c_3 \end{vmatrix} + c_1 \begin{vmatrix} a_2 & b_2 \\ a_3 & b_3 \end{vmatrix}$$

$$= a_1 (b_2c_3 - b_3c_2) - b_1 (a_2c_3 - a_3c_2) + c_1 (a_2b_3 - a_3b_2)$$

A square matrix is called a *diagonal matrix* if all the elements other than those along the diagonal are zero. For example,

$$C = \begin{pmatrix} b_1 & 0 & 0 \\ 0 & b_2 & 0 \\ 0 & 0 & b_3 \end{pmatrix}$$

If every element of a diagonal matrix is 1, it is known as a *unit matrix, E*. For a matrix A, there exists a matrix A^{-1} (inverse of A) such that $AA^{-1} = A^{-1}A = E$. The inverse does not exist if $|A| = 0$. Such matrices are called *singular matrices*. A and A^{-1} commute.

3.1.1 Transpose of a Matrix

The *transpose* of a matrix is obtained by replacing every element a_{ij} with element a_{ji}.

$$\text{Consider a matrix } A = \begin{pmatrix} a_{11} & a_{12} & a_{13} \\ b_{21} & b_{22} & b_{23} \\ c_{31} & c_{32} & c_{33} \end{pmatrix}$$

This is a 3 × 3 matrix. The transpose of this matrix is given by

$$A^t = \begin{pmatrix} a_{11} & b_{21} & c_{31} \\ a_{12} & b_{22} & c_{32} \\ a_{13} & b_{23} & c_{33} \end{pmatrix} \text{ which is also a } 3 \times 3 \text{ matrix}$$

Note: If a matrix is of the order n× k (where n = number of rows and k = number of columns), its transpose is of an order k × n.

3.1.2 Minors and Cofactors

Minor: The *minor* of an element in a matrix is the determinant obtained by suppressing the row and column in which the element appears. The order of the minor of an element in the matrix A is one less than the order of the matrix A.

The minor of an element a_{ij} is denoted by M_{ij}.

Cofactor: The minor with the appropriate sign is called the *cofactor*. The cofactor of an element a_{ij} is denoted by $C_{ij} = (-1)^{i+j} M_{ij}$.

3.1.3 Adjoint of a Matrix

The *adjoint* of a square matrix A is the transpose of the matrix formed by the cofactors of the elements of determinant $|A|$. It is denoted by *adj A*.

Let $A = \begin{pmatrix} a_{11} & a_{12} & a_{13} \\ a_{21} & a_{22} & a_{23} \\ a_{31} & a_{32} & a_{33} \end{pmatrix}$ and let A_{ij} denote the cofactor of the element $\mathbf{a_{ij}}$ in the determinant $|A|$. The matrix formed by the cofactors of elements of determinant

$$|A| \text{ is} = \begin{pmatrix} A_{11} & A_{12} & A_{13} \\ A_{21} & A_{22} & A_{23} \\ A_{31} & A_{32} & A_{33} \end{pmatrix}$$

$$\therefore\ adj\ A = \begin{pmatrix} A_{11} & A_{21} & A_{31} \\ A_{12} & A_{22} & A_{32} \\ A_{13} & A_{23} & A_{33} \end{pmatrix}$$

3.1.4 Inverse of a Matrix

A square matrix A is said to be a singular matrix if the value of the determinant $|A|$ is zero. If the value of the determinant is non-zero, then the square matrix A is called a non-singular matrix.

Let A be a non-singular square matrix. Then, there always exists a square matrix B such that $BA = AB = 1$ (unit matrix having the same order as that of A) and B is called the inverse matrix of A and is denoted by A^{-1}.

If A is a non-singular square matrix, then the inverse of A is

$$A^{-1} = \frac{1}{|A|}\ adj\ A$$

Example 3.1: Find the inverse of the matrix $A = \begin{pmatrix} 1 & 1 & 1 \\ 4 & 3 & 1 \\ 1 & 5 & 2 \end{pmatrix}$

$$|A| = \begin{vmatrix} 1 & 1 & 1 \\ 4 & 3 & 1 \\ 1 & 5 & 2 \end{vmatrix}$$

$$= 1(6-5) - 1(8-1) + 1(20-3)$$

$$= 1 - 7 + 17 = 11$$

$\because \quad A \neq 0, A^{-1}$ exists.

Cofactors of the matrix elements are

$$A_{11} = +\begin{vmatrix} 3 & 1 \\ 5 & 2 \end{vmatrix} = 6 - 5 = 1$$

$$A_{12} = -\begin{vmatrix} 4 & 1 \\ 1 & 2 \end{vmatrix} = -(8 - 1) = -7$$

$$A_{13} = +\begin{vmatrix} 4 & 3 \\ 1 & 5 \end{vmatrix} = 20 - 3 = 17$$

$$A_{21} = -\begin{vmatrix} 1 & 1 \\ 5 & 2 \end{vmatrix} = -(2 - 5) = 3$$

$$A_{22} = +\begin{vmatrix} 1 & 1 \\ 1 & 2 \end{vmatrix} = 2 - 1 = 1$$

$$A_{23} = -\begin{vmatrix} 1 & 1 \\ 1 & 5 \end{vmatrix} = -(5 - 1) = -4$$

$$A_{31} = +\begin{vmatrix} 1 & 1 \\ 3 & 1 \end{vmatrix} = 1 - 3 = -2$$

$$A_{32} = -\begin{vmatrix} 1 & 1 \\ 4 & 1 \end{vmatrix} = -(1 - 4) = 3$$

$$A_{33} = +\begin{vmatrix} 1 & 1 \\ 4 & 3 \end{vmatrix} = 3 - 4 = -1$$

$$\therefore adj\, A = \begin{pmatrix} A_{11} & A_{12} & A_{13} \\ A_{21} & A_{22} & A_{23} \\ A_{31} & A_{32} & A_{33} \end{pmatrix}^t$$

$$= \begin{bmatrix} 1 & -7 & 17 \\ 3 & 1 & -4 \\ -2 & 3 & -1 \end{bmatrix}^t = \begin{bmatrix} 1 & 3 & -2 \\ -7 & 1 & 3 \\ 17 & -4 & -1 \end{bmatrix}$$

$$A^{-1} = \frac{adj\, A}{|A|}$$

$$= 1/11 \begin{bmatrix} 1 & 3 & -2 \\ -7 & 1 & 3 \\ 17 & -4 & -1 \end{bmatrix} = \begin{bmatrix} 1/11 & 3/11 & -2/11 \\ -7/11 & 1/11 & 3/11 \\ 17/11 & -4/11 & -1/11 \end{bmatrix}$$

3.2 Solutions of Linear Equations by Matrix Method

Let the linear equations in x, y and z be

$$\begin{aligned} a_1x + b_1y + c_1z &= d_1 \\ a_2x + b_2y + c_2z &= d_2 \\ a_3x + b_3y + c_3z &= d_3 \end{aligned}$$

$$\text{Let } A = \begin{pmatrix} a_1 & b_1 & c_1 \\ a_2 & b_2 & c_2 \\ a_3 & b_3 & c_3 \end{pmatrix} \qquad X = \begin{pmatrix} x \\ y \\ z \end{pmatrix} \qquad \text{and } B = \begin{pmatrix} d_1 \\ d_2 \\ d_3 \end{pmatrix}$$

The linear equations can be combined to give the following matrix equation.

$$\begin{pmatrix} a_1 & b_1 & c_1 \\ a_2 & b_2 & c_2 \\ a_3 & b_3 & c_3 \end{pmatrix} \begin{pmatrix} x \\ y \\ z \end{pmatrix} = \begin{pmatrix} d_1 \\ d_2 \\ d_3 \end{pmatrix}$$

That is, $A\,X = B$

If A is a non-singular square matrix, A^{-1} exists and hence

$$A^{-1}(A\,X) = A^{-1}\,B$$

That is, $(A^{-1}A)\,X = A^{-1}\,B$

or $\quad E\,X = A^{-1}B$

$\therefore \quad X = A^{-1}\,B$

Because A is a non-singular matrix, A^{-1} is also a 3×3 matrix. B is a 3×1 matrix and hence $A^{-1}B$ will be a 3×1 matrix.

$$\begin{pmatrix} x \\ y \\ z \end{pmatrix} = A^{-1} \begin{pmatrix} d_1 \\ d_2 \\ d_3 \end{pmatrix} = A^{-1}d$$

Equating the corresponding elements on both sides, the solution can be obtained.

That is, $x = (A^{-1}\,d)_{11}$

$y = (A^{-1}\,d)_{21}$

$z = (A^{-1}\,d)_{31}$

3.3 Transformation Matrices for Symmetry Elements

The changes produced by symmetry operations on a molecule have been illustrated by figures in Chapter 2. If the configurations are not carefully examined, it may lead to ambiguous conclusions. The most accurate method is to study the effect of operations on the coordinates and express the transformations in the form of a matrix. Such matrices are referred to as transformation matrices, since they indicate the net quantitative effect of the symmetry operations.

3.3.1 Transformation Matrix for Identity Element

Consider a vector $\overrightarrow{OP}$ having the coordinates (x_1, y_1, z_1). When the identity operation '*E*' is performed on this vector,let the coordinates be changed to (x_2, y_2, z_2).

$$\overrightarrow{OP}(x_1, y_1, z_1) \xrightarrow{E} \overrightarrow{OP}(x_2, y_2, z_2)$$

Since by definition, the identity operation should not change any coordinates, $x_2 = x_1$, $y_2 = y_1$ and $z_2 = z_1$.

The new position of the vector can be defined in terms of the original coordinates of the vector and the relationship between (x_2, y_2, z_2) and (x_1, y_1, z_1) can be expressed as a set of mathematical equations.

$$\begin{aligned} x_2 &= (1 \times x_1) + (0 \times y_1) + (0 \times z_1) \\ y_2 &= (0 \times x_1) + (1 \times y_1) + (0 \times z_1) \\ z_2 &= (0 \times x_1) + (0 \times y_1) + (1 \times z_1) \end{aligned}$$

These equations can be combined to give the following matrix equation.

$$\begin{pmatrix} x_2 \\ y_2 \\ z_2 \end{pmatrix} = \begin{pmatrix} 1 & 0 & 0 \\ 0 & 1 & 0 \\ 0 & 0 & 1 \end{pmatrix} \begin{pmatrix} x_1 \\ y_1 \\ z_1 \end{pmatrix}$$

The matrix $\begin{pmatrix} 1 & 0 & 0 \\ 0 & 1 & 0 \\ 0 & 0 & 1 \end{pmatrix}$ is the transformation matrix for *E* since it indicates the net quantitative effect of E on the vector. A 3 × 3 matrix is obtained since we started with three components (also called the basis elements) of a vector.

The above procedure gives an idea about the changes that have taken place with respect to the original coordinates that results in new coordinates.

The character for the transformation matrix *E*,

χ_E = Sum of diagonal elements (from left top to right bottom)
= 3.

3.3.2 Transformation Matrix for Plane of Symmetry (Reflection)

Let σ_{xy} be performed on a vector $\overrightarrow{OP}(x_1, y_1, z_1)$. The original coordinates (x_1, y_1, z_1) get transformed to $(x_2, y_2, -z_2)$.

$$\overrightarrow{OP}(x_1, y_1, z_1) \xrightarrow{\sigma_{xy}} \overrightarrow{OP}(x_2, y_2, -z_2)$$

During reflection along the xy plane, only the z coordinate gets inverted.

That is, $x_2 = x_1$
$y_2 = y_1$
$z_2 = -z_1$

The new position of the vector is defined by the combination of the original coordinates and this generates a set of mathematical equations.

$$x_2 = (1 \times x_1) + (0 \times y_1) + (0 \times z_1)$$
$$y_2 = (0 \times x_1) + (1 \times y_1) + (0 \times z_1)$$
$$z_2 = (0 \times x_1) + (0 \times y_1) + (-1 \times z_1)$$

These simultaneous equations can be combined to give the following matrix equation.

$$\begin{pmatrix} x_2 \\ y_2 \\ z_2 \end{pmatrix} = \begin{pmatrix} 1 & 0 & 0 \\ 0 & 1 & 0 \\ 0 & 0 & -1 \end{pmatrix} \begin{pmatrix} x_1 \\ y_1 \\ z_1 \end{pmatrix}$$

The matrix $\begin{pmatrix} x_2 \\ y_2 \\ z_2 \end{pmatrix}$ represents the position corresponding to the new coordinates of the vector.

The matrix $\begin{pmatrix} x_1 \\ y_1 \\ z_1 \end{pmatrix}$ represents the position corresponding to the original coordinates of the vector and $\begin{pmatrix} 1 & 0 & 0 \\ 0 & 1 & 0 \\ 0 & 0 & -1 \end{pmatrix}$ corresponds to the transformation matrix for the reflection σ_{xy}.

$$\chi_{\sigma xy} = 1$$

Similarly, the transformation matrix for $\sigma_{xz} = \begin{pmatrix} 1 & 0 & 0 \\ 0 & -1 & 0 \\ 0 & 0 & 1 \end{pmatrix}$

and $\chi_{\sigma xz} = 1$,

and, the transformation matrix for $\sigma_{yz} = \begin{pmatrix} -1 & 0 & 0 \\ 0 & 1 & 0 \\ 0 & 0 & 1 \end{pmatrix}$ and $\chi_{\sigma yz} = 1$.

3.3.3 Transformation Matrix for Centre of Symmetry

When inversion operation is performed on a vector $\overrightarrow{OP}(x_1, y_1, z_1)$, it changes to the vector with coordinates (x_2, y_2, z_2).

$$\overrightarrow{OP}(x_1, y_1, z_1) \xrightarrow{i} \overrightarrow{OP}(-x_1, -y_1, -z_1) = \overrightarrow{OP}(x_2, y_2, z_2)$$

$$\therefore \quad x_2 = -x_1$$
$$y_2 = -y_1$$
$$\text{and} \quad z_2 = -z_1$$

The relationship between (x_1, y_1, z_1) and (x_2, y_2, z_2) can be represented by a set of mathematical equations.

$$x_2 = (-1) \times x_1 + 0 \times y_1 + 0 \times z_1$$
$$y_2 = 0 \times x_1 + (-1) \times y_1 + 0 \times z_1$$
$$z_2 = 0 \times x_1 + 0 \times y_1 + (-1) \times z_1$$

The corresponding matrix equation can be given as,

$$\begin{pmatrix} x_2 \\ y_2 \\ z_2 \end{pmatrix} = \begin{pmatrix} -1 & 0 & 0 \\ 0 & -1 & 0 \\ 0 & 0 & -1 \end{pmatrix} \begin{pmatrix} x_1 \\ y_1 \\ z_1 \end{pmatrix}$$

$\therefore$ The transformation matrix for $i = \begin{pmatrix} -1 & 0 & 0 \\ 0 & -1 & 0 \\ 0 & 0 & -1 \end{pmatrix}$

and $\chi_i = -3$.

3.3.4 Transformation Matrix for Axis of Symmetry

Consider Fig. 3.1. Let a vector $\overrightarrow{OP}$ have a magnitude 'r' and derive an angle θ with respect to the x axis. Assume that $\overrightarrow{OP}$ is rotated about the z axis in the clockwise direction through an angle of ϕ. The new vector $\overrightarrow{OP'}$ derives an angle $(\theta - \phi)$ with the x axis. The z coordinates of the points do not change and $z_2 = z_1$.

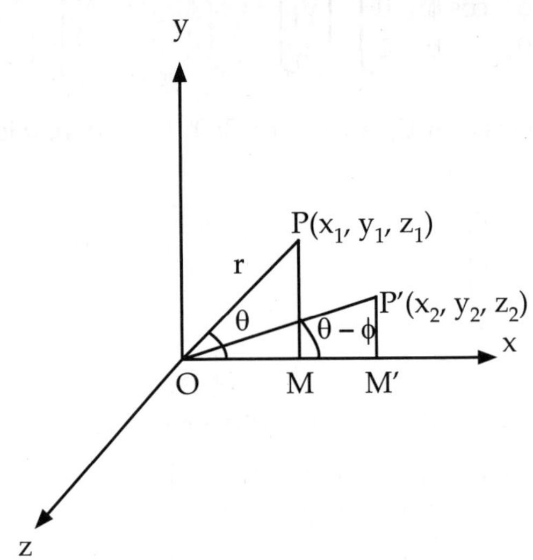

Fig 3.1

From the figure, we can see that,

$$OM = x_1 = r\cos\theta \qquad OM' = x_2 = r\cos(\theta-\phi)$$
$$PM = y_1 = r\sin\theta \qquad P'M' = y_2 = r\sin(\theta-\phi)$$

Using the identities,

$$\cos(A-B) = \cos A\cos B + \sin A\sin B$$
$$\sin(A-B) = \sin A\cos B - \cos A\cos B$$

the relationship between (x_2, y_2, z_2) and (x_1, y_1, z_1) can be expressed as,

$$x_2 = r\cos\theta\cos\phi + r\sin\theta\sin\phi$$
$$y_2 = r\sin\theta\cos\phi - r\cos\theta\sin\phi$$
$$z_2 = z_1$$

Substituting for $r\cos\theta$ and $r\sin\theta$ as x_1 and y_1, we obtain,

$$x_2 = x_1\cos\phi + y_1\sin\phi$$
$$y_2 = -x_1\cos\phi + y_1\sin\phi$$
$$z_2 = z_1$$

Then, the new position of the vector $\overrightarrow{OP}$ can be expressed in terms of the original coordinates as,

$$x_2 = x_1\cos\phi + y_1\sin\phi + 0\times z_1$$
$$y_2 = -x_1\sin\phi + y_1\cos\phi + 0\times z_1$$
$$z_2 = 0\times x_1 + 0\times y_1 + 1\times z_1$$

These three simultaneous equations can be expressed in terms of a matrix equation as follows,

$$\begin{pmatrix} x_2 \\ y_2 \\ z_2 \end{pmatrix} = \begin{pmatrix} \cos\phi & \sin\phi & 0 \\ -\sin\phi & \cos\phi & 0 \\ 0 & 0 & 1 \end{pmatrix} \begin{pmatrix} x_1 \\ y_1 \\ z_1 \end{pmatrix}$$

The transformation matrix for C_n where $\phi = 360°/n$ (where n is the order of the axis) is given by,

$$\begin{pmatrix} \cos\phi & \sin\phi & 0 \\ -\sin\phi & \cos\phi & 0 \\ 0 & 0 & 1 \end{pmatrix}$$

and $\chi_{C_{nh}} = 2\cos\phi + 1$

When $n = 2$, $\phi = 360°/2 = 180°$ and the corresponding 3×3 matrix is given by,

$$\begin{pmatrix} \cos 180 & \sin 180 & 0 \\ -\sin 180 & \cos 180 & 0 \\ 0 & 0 & 1 \end{pmatrix} = \begin{pmatrix} -1 & 0 & 0 \\ 0 & -1 & 0 \\ 0 & 0 & 1 \end{pmatrix}$$

and $\chi_{C_2} = 1$.

When $n = 3$, $\phi = 360° / 3 = 120°$, the 3×3 transformation matrix is given by,

$$\begin{pmatrix} \cos 120 & \sin 120 & 0 \\ -\sin 120 & \cos 120 & 0 \\ 0 & 0 & 1 \end{pmatrix} = \begin{pmatrix} -1/2 & \sqrt{3}/2 & 0 \\ \sqrt{3}/2 & -1/2 & 0 \\ 0 & 0 & 1 \end{pmatrix}$$

and $\chi_{C_3} = 0$.

When $n = 4$, $\phi = 360° / 4 = 90°$, the 3×3 transformation matrix given by,

$$\begin{pmatrix} \cos 90 & \sin 90 & 0 \\ -\sin 90 & \cos 90 & 0 \\ 0 & 0 & 1 \end{pmatrix} = \begin{pmatrix} 0 & 1 & 0 \\ -1 & 0 & 0 \\ 0 & 0 & 1 \end{pmatrix}$$

$\therefore \chi_{C_4} = 1$.

3.4.5 Transformation Matrix for Improper Axis of Symmetry

The S_n operation involves rotation about an axis followed by reflection about a plane perpendicular to that axis. The transformation matrix for S_n can be obtained by finding the product of the transformation matrices of $C_n(z)$ and σ_{xy}.

$$S_n = \sigma_{xy} \; C_{n(z)}$$

$$\text{Transformation matrix of } S_n = \begin{pmatrix} 1 & 0 & 0 \\ 0 & 1 & 0 \\ 0 & 0 & -1 \end{pmatrix} \begin{pmatrix} \cos\phi & \sin\phi & 0 \\ -\sin\phi & \cos\phi & 0 \\ 0 & 0 & 1 \end{pmatrix}$$

$$= \begin{pmatrix} \cos\phi & \sin\phi & 0 \\ -\sin\phi & \cos\phi & 0 \\ 0 & 0 & -1 \end{pmatrix}$$

and $\chi_{S_n} = 2\cos\phi - 1$.

Thus, this chapter has highlighted how various symmetry operations can be represented using matrices. The products of symmetry operations can be determined by the multiplication of the respective matrices. For example,

$$\sigma_{yz} \times \sigma_{xz} = \begin{pmatrix} -1 & 0 & 0 \\ 0 & 1 & 1 \\ 0 & 0 & 1 \end{pmatrix} \times \begin{pmatrix} 1 & 0 & 0 \\ 0 & -1 & 0 \\ 0 & 0 & 1 \end{pmatrix} = \begin{pmatrix} -1 & 0 & 0 \\ 0 & -1 & 0 \\ 0 & 0 & 1 \end{pmatrix} = C_{2z}$$

4

Representation of Point Groups

4.1 Matrix Representation

As we have seen in Chapter 3, the best way to understand how the symmetry operations of a molecule influence its properties, is to study the sets of matrices generated by various symmetry operations of that particular point group to which the molecule belongs. A collection of such matrices corresponding to every symmetry element present in a point group is referred to as the matrix representation of a point group. Thus, matrix representations replace the geometry of symmetry operations with the algebra of matrices. These representations are the crucial link between the symmetry of molecules and the practical applications of group theory in chemistry.

4.2 Characteristics of Matrix Representations

Matrix representations of point groups should obey the following mathematical rules.

(i) The product of any two matrices and the square of any matrix belonging to the point group, should also be a matrix that belongs to the point group.
(ii) There should be one matrix, which when combined with any other matrix, leaves that matrix unchanged.
(iii) Associative law should hold good.
(iv) The inverse of every matrix of the point group should also be present in the representation for that point group.

Thus, the collection of matrices corresponding to each symmetry element of a point group which obeys all the four mathematical rules for the formation of a group, and produces a group multiplication table identical to that derived from the symmetry operations is called a matrix representation of a point group.

4.3 CHARACTER REPRESENTATION OF A POINT GROUP

The *trace* of a transformation matrix that represents either a symmetry element or symmetry operation of a group is referred to as the *character* of the matrix. The trace or character of the matrix is the sum of its diagonal elements. The character of the matrix is represented by the symbol χ. Thus $\chi_{(R)}$ represents the character of the symmetry operation R in the matrix representation.

$$\therefore \chi_{(R)} = \sum_{i} A_{ii}$$

Where, A is the matrix element belonging to the ith row and the ith column.

The collection of characters for every symmetry element of a group is called the character representation. Such character representations should also obey the mathematical rules for the formation of a group. It has also been mathematically proven that if the characters of two representations are identical, then the two representations are necessarily equivalent.

4.4 DERIVATION OF REPRESENTATIONS

There are several methods for obtaining sets of transformation matrices that are isomorphic with a given point group. One method is to consider the effect of symmetry operations of a point group on the cartesian coordinates of some point, or some position vector in the molecule. Another method is to consider the effect of symmetry operations on one or more sets of base vectors within the molecule. Thus, by choosing different position vectors and base vectors, a large number of matrix representations can be derived for a group. The base vectors chosen are said to form the base for the representation of the group. Such matrix representations or character representations can thus be obtained by choosing appropriate base vectors and systematically deriving the transformation matrices for every symmetry operation of a point group. They can also be derived by simple methods.

4.4.1 Systematic Method for Deriving Matrix Representations

It is apparent that we can get a set of 3 × 3 matrices which form a representation of a given point group by considering the effect of symmetry operations of a point group on the three cartesian coordinates (see Chapter 2). Such a 3 × 3 representation can be given for the C_{2v} point group as follows.

$$C_{2v} \quad \underset{E}{\begin{pmatrix} 1 & 0 & 0 \\ 0 & 1 & 0 \\ 0 & 0 & 1 \end{pmatrix}} \quad \underset{C_{2(Z)}}{\begin{pmatrix} -1 & 0 & 0 \\ 0 & -1 & 0 \\ 0 & 0 & 1 \end{pmatrix}} \quad \underset{\sigma_{xz}}{\begin{pmatrix} 1 & 0 & 0 \\ 0 & -1 & 0 \\ 0 & 0 & 1 \end{pmatrix}} \quad \underset{\sigma_{yz}}{\begin{pmatrix} -1 & 0 & 0 \\ 0 & 1 & 0 \\ 0 & 0 & 1 \end{pmatrix}}$$

We can also derive a 4×4 matrix representation for the same C_{2v} point group by considering a different basis set for the representation.

Let us consider the example of 1,3-butadiene which is a planar molecule. This molecule belongs to the C_{2v} point group and the various symmetry elements of the group are E, $C_{2(z)}$, σ_{xz} and σ_{yz}. Each carbon atom in 1,3-butadiene has an unhybridised p orbital and the four unhybridised p_Z orbitals are chosen as the bases for the representation. Let ϕ_1, ϕ_2, ϕ_3 and ϕ_4 be the original configurations of the p orbitals and ϕ_1', ϕ_2', ϕ_3' and ϕ_4' be the new positions of the orbitals after performing the symmetry operations.

(i) After performing the identity operation (E), we know that

$$\phi_1' = \phi_1,\ \phi_2' = \phi_2,\quad \phi_3' = \phi_3,\ \phi_4' = \phi_4$$

ϕ_1', ϕ_2', ϕ_3', ϕ_4' can be defined in terms of the original positions of the orbitals as follows.

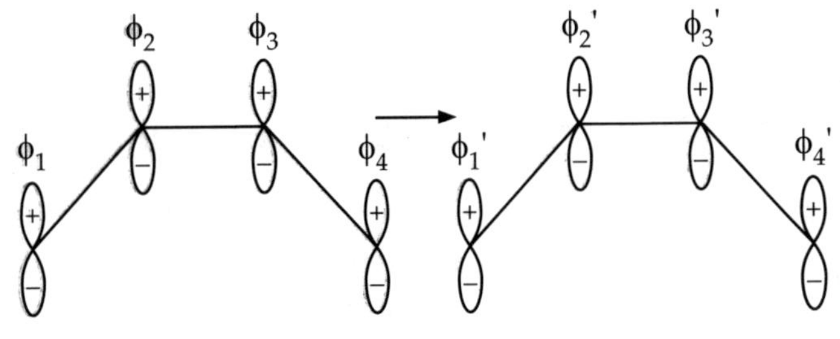

$$\begin{pmatrix} \phi_1' \\ \phi_2' \\ \phi_3' \\ \phi_4' \end{pmatrix} = \begin{pmatrix} 1 & 0 & 0 & 0 \\ 0 & 1 & 0 & 0 \\ 0 & 0 & 1 & 0 \\ 0 & 0 & 0 & 1 \end{pmatrix} \begin{pmatrix} \phi_1 \\ \phi_2 \\ \phi_3 \\ \phi_4 \end{pmatrix}$$

Thus, the transformation matrix corresponding to the symmetry element E for the unhybridised p_Z orbitals of the 1,3-butadiene molecule is

$$\begin{pmatrix} 1 & 0 & 0 & 0 \\ 0 & 1 & 0 & 0 \\ 0 & 0 & 1 & 0 \\ 0 & 0 & 0 & 1 \end{pmatrix} \text{ and } \therefore\ \chi_{(E)} = 4.$$

(ii) When the $C_{2(z)}$ operation is performed on unhybridised orbitals, it is known that,

$$\phi_1' = -\phi_4,\ \phi_2' = -\phi_3,\ \phi_3' = -\phi_2,\ \phi_4' = -\phi_1$$

Defining new configurations of the unhybridised orbitals in terms of their original configurations, we get,

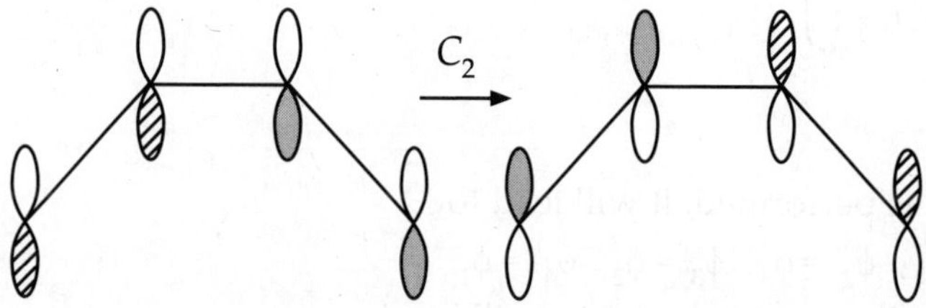

$$\begin{pmatrix} \phi_1' \\ \phi_2' \\ \phi_3' \\ \phi_4' \end{pmatrix} = \begin{pmatrix} 0 & 0 & 0 & -1 \\ 0 & 0 & -1 & 0 \\ 0 & -1 & 0 & 0 \\ -1 & 0 & 0 & 0 \end{pmatrix} \begin{pmatrix} \phi_1 \\ \phi_2 \\ \phi_3 \\ \phi_4 \end{pmatrix}$$

Therefore, the transformation matrix for C_{2z} is,

$$\begin{pmatrix} 0 & 0 & 0 & -1 \\ 0 & 0 & -1 & 0 \\ 0 & -1 & 0 & 0 \\ -1 & 0 & 0 & 0 \end{pmatrix}$$

and its corresponding character is zero. $\chi_{C_{2(z)}} = 0$.

(iii) When σ_{xz} is performed (reflection about the molecular plane),

$$\phi_1' = -\phi_1,\ \phi_2' = -\phi_2,\ \phi_3' = -\phi_3,\ \phi_4' = -\phi_4$$

Expressing the new positions of the *p* orbitals in terms of their original positions, we have,

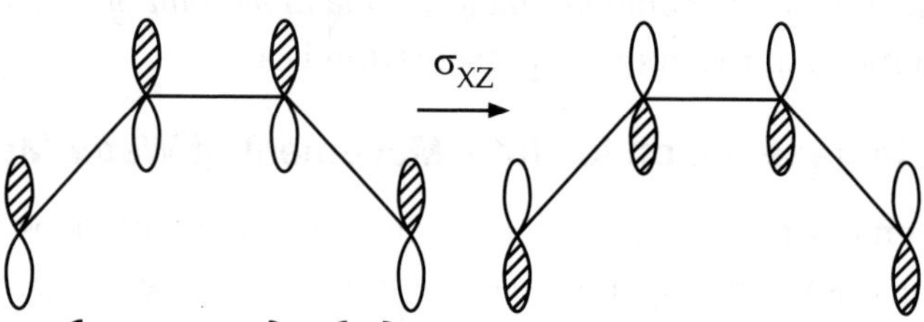

$$\begin{pmatrix} \phi_1' \\ \phi_2' \\ \phi_3' \\ \phi_4' \end{pmatrix} = \begin{pmatrix} -1 & 0 & 0 & 0 \\ 0 & -1 & 0 & 0 \\ 0 & 0 & -1 & 0 \\ 0 & 0 & 0 & -1 \end{pmatrix} \begin{pmatrix} \phi_1 \\ \phi_2 \\ \phi_3 \\ \phi_4 \end{pmatrix}$$

Therefore, the transformation matrix for σ_{xz} is

$$\begin{pmatrix} -1 & 0 & 0 & 0 \\ 0 & -1 & 0 & 0 \\ 0 & 0 & -1 & 0 \\ 0 & 0 & 0 & -1 \end{pmatrix}$$

and $\chi_{\sigma_{xz}} = -4$.

(iv) When σ_{yz} is performed, it will lead to

$$\phi_1' = \phi_4,\ \phi_2' = \phi_3,\ \phi_3' = \phi_2,\ \phi_4' = \phi_1$$

The corresponding matrix equation will be given as,

$$\begin{pmatrix} \phi_1' \\ \phi_2' \\ \phi_3' \\ \phi_4' \end{pmatrix} = \begin{pmatrix} 0 & 0 & 0 & 1 \\ 0 & 0 & 1 & 0 \\ 0 & 1 & 0 & 0 \\ 1 & 0 & 0 & 0 \end{pmatrix} \begin{pmatrix} \phi_1 \\ \phi_2 \\ \phi_3 \\ \phi_4 \end{pmatrix}$$

The transformation matrix for σ_{yz} is

$$\begin{pmatrix} 0 & 0 & 0 & -1 \\ 0 & 0 & -1 & 0 \\ 0 & -1 & 0 & 0 \\ -1 & 0 & 0 & 0 \end{pmatrix} \quad \text{and } \chi_{\sigma_{yz}} = 0.$$

In the above example of butadiene, the characters corresponding to the various symmetry elements for the unhybridised *p* orbitals are

	E	$C_{2(z)}$	σ_{xz}	σ_{yz}
$\chi_{(R)}$	4	0	–4	0

This is called the character representation. *The character corresponding to the identity element in any representation indicates the dimension of that representation.* Thus, the dimension of the above representation is 4.

4.4.2 Matrix Representation for Total Movement of Water Molecule

The total movement of the atoms in the water molecule involves three vectors as each atom is associated with one vector. There are nine components. These vectors form the base of the representation.

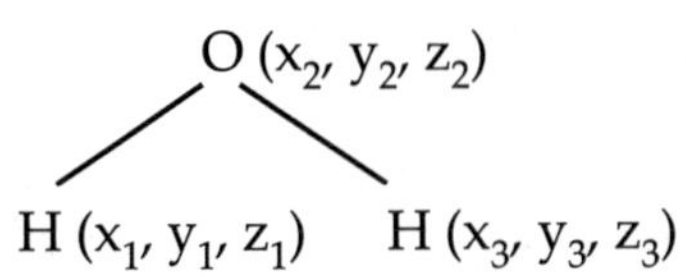

Let (x_1, y_1, z_1), (x_2, y_2, z_2) and (x_3, y_3, z_3) be the original positions of the vectors associated with the atoms of the water molecule. After performing the symmetry operations, let (x_1', y_1', z_1'), (x_2', y_2', z_2') and (x_3', y_3', z_3') represent the new positions of the vectors.

(i) When the symmetry operation E is performed,

$x_1' = x_1,\ y_1' = y_1,\ z_1' = z_1,\ x_2' = x_2,\ y_2' = y_2,\ z_2' = z_2,\ x_3' = x_3,\ y_3' = y_3,\ z_3' = z_3$

Expressing the new positions of the vectors in terms of their original positions and combining the equations into a matrix equation leads to the following transformation matrix.

$$\left(\begin{array}{ccc|ccc|ccc} 1 & 0 & 0 & 0 & 0 & 0 & 0 & 0 & 0 \\ 0 & 1 & 0 & 0 & 0 & 0 & 0 & 0 & 0 \\ 0 & 0 & 1 & 0 & 0 & 0 & 0 & 0 & 0 \\ \hline 0 & 0 & 0 & 1 & 0 & 0 & 0 & 0 & 0 \\ 0 & 0 & 0 & 0 & 1 & 0 & 0 & 0 & 0 \\ 0 & 0 & 0 & 0 & 0 & 1 & 0 & 0 & 0 \\ \hline 0 & 0 & 0 & 0 & 0 & 0 & 1 & 0 & 0 \\ 0 & 0 & 0 & 0 & 0 & 0 & 0 & 1 & 0 \\ 0 & 0 & 0 & 0 & 0 & 0 & 0 & 0 & 1 \end{array}\right)$$

and $\chi_E = 9$.

(ii) When the C_{2z} operation is performed,

$x_1' = -x_3,\ y_1' = -y_3,\ z_1' = z_3,\ x_2' = -x_2,\ y_2' = -y_2,\ z_2' = z_2,\ x_3' = -x_1,\ y_3' = -y_1,\ z_3' = z_1$

Expressing the new positions of the vectors in terms of their original positions, and combining the equations into a matrix equation leads to the following transformation matrix.

$$\left(\begin{array}{ccc|ccc|ccc} 0 & 0 & 0 & 0 & 0 & 0 & -1 & 0 & 0 \\ 0 & 0 & 0 & 0 & 0 & 0 & 0 & -1 & 0 \\ 0 & 0 & 0 & 0 & 0 & 0 & 0 & 0 & 1 \\ \hline 0 & 0 & 0 & -1 & 0 & 0 & 0 & 0 & 0 \\ 0 & 0 & 0 & 0 & -1 & 0 & 0 & 0 & 0 \\ 0 & 0 & 0 & 0 & 0 & 1 & 0 & 0 & 0 \\ \hline -1 & 0 & 0 & 0 & 0 & 0 & 0 & 0 & 0 \\ 0 & -1 & 0 & 0 & 0 & 0 & 0 & 0 & 0 \\ 0 & 0 & 1 & 0 & 0 & 0 & 0 & 0 & 0 \end{array}\right)$$

$\therefore\ \chi_{C_{2(z)}} = -1 - 1 + 1 = -1$

(iii) When the σ_{xz} is performed (molecular plane reflection),

$x_1' = x_1,\ y_1' = -y_1,\ z_1' = z_1,\ x_2' = x_2,\ y_2' = -y_2,\ z_2' = z_2,\ x_3' = x_3,\ y_3' = -y_3,\ z_3' = z_3$

Expressing the new positions of the vectors in terms of their original positions, and combining the equations into a matrix equation leads to the following transformation matrix.

$$\left(\begin{array}{ccc|ccc|ccc} 1 & 0 & 0 & 0 & 0 & 0 & 0 & 0 & 0 \\ 0 & -1 & 0 & 0 & 0 & 0 & 0 & 0 & 0 \\ 0 & 0 & 1 & 0 & 0 & 0 & 0 & 0 & 0 \\ \hline 0 & 0 & 0 & 1 & 0 & 0 & 0 & 0 & 0 \\ 0 & 0 & 0 & 0 & -1 & 0 & 0 & 0 & 0 \\ 0 & 0 & 0 & 0 & 0 & 1 & 0 & 0 & 0 \\ \hline 0 & 0 & 0 & 0 & 0 & 0 & 1 & 0 & 0 \\ 0 & 0 & 0 & 0 & 0 & 0 & 0 & -1 & 0 \\ 0 & 0 & 0 & 0 & 0 & 0 & 0 & 0 & 1 \end{array}\right)$$

$$\chi_{\sigma_{xz}} = 6 - 3 = 3$$

(iv) When the σ_{yz} operation is performed,

$x_1' = -x_3,\ y_1' = y_3,\ z_1' = z_3,\ x_2' = -x_2,\ y_2' = y_2,\ z_2' = z_2,\ x_3' = -x_1,\ y_3' = y_1,\ z_3' = z_1$

Expressing the new positions of the vectors in terms of their original positions, and combining the equations into a matrix equation leads to the following transformation matrix.

$$\left(\begin{array}{ccc|ccc|ccc} 0 & 0 & 0 & 0 & 0 & 0 & -1 & 0 & 0 \\ 0 & 0 & 0 & 0 & 0 & 0 & 0 & 1 & 0 \\ 0 & 0 & 0 & 0 & 0 & 0 & 0 & 0 & 1 \\ \hline 0 & 0 & 0 & -1 & 0 & 0 & 0 & 0 & 0 \\ 0 & 0 & 0 & 0 & 1 & 0 & 0 & 0 & 0 \\ 0 & 0 & 0 & 0 & 0 & 1 & 0 & 0 & 0 \\ \hline -1 & 0 & 0 & 0 & 0 & 0 & 0 & 0 & 0 \\ 0 & 1 & 0 & 0 & 0 & 0 & 0 & 0 & 0 \\ 0 & 0 & 1 & 0 & 0 & 0 & 0 & 0 & 0 \end{array}\right)$$

$$\chi_{\sigma_{yz}} = -1 + 1 + 1 = +1$$

The corresponding character representation of the molecule is given by,

	E	C_{2z}	σ_{xz}	σ_{yz}
χ	9	–1	3	1

The dimension of this representation is 9.

4.4.3 Simple Method to Directly Arrive at the Character Representation of a Point Group

Method I: The character corresponding to the symmetry operation R for the given base vectors or given sets of vectors, can be derived from the following equation.

$$\chi_R = \begin{array}{l}\text{No. of vectors or no. of orbitals}\\ \text{which do not change position}\\ \text{and direction when a symmetry}\\ \text{operation (R) is performed}\end{array} - \begin{array}{l}\text{No. of vectors or orbitals which}\\ \text{do not shift but simply reverse}\\ \text{direction when a symmetry}\\ \text{operation (R) is performed}\end{array}$$

This method can be applied to derive the character representation for the unhybridised *p* orbitals of *cis*- 1,3-butadiene which belongs to the C_{2v} point group. Accordingly,

$\chi_E = 4 - 0 = 4$

$\chi_{C_{2(z)}} = 0 - 0 = 0$

$\chi_{\sigma_{xz}} = 0 - 4 = -4$

(Molecular plane)

$\chi_{\sigma_{yz}} = 0 - 0 = 0$

Hence, the representation is 4 0 –4 0 which is similar to the one derived from the systematic method. Similarly, the character representation for the total movement of the water molecule can also be derived using the same method.

$\chi_E = 9 - 0 = 9$

$\chi_{C_{2(z)}} = 1 - 2 = -1$

$\chi_{\sigma_{xz}} = 6 - 3 = 3$

(Molecular plane)

$\chi_{\sigma_{yz}} = 2 - 1 = 1$

Thus, the representation is 9 –1 3 1.

Method II: In the second method, the character corresponding to any symmetry operation *R* in a representation is given by,

$$\chi_R = \begin{array}{l}\text{No. of atoms that do not shift}\\ \text{when a symmetry operation}\\ (R) \text{ is performed}\end{array} \times \begin{array}{l}\text{Character corresponding to the}\\ \text{symmetry operation } (R) \text{ in a general}\\ (3 \times 3) \text{ matrix representation}\end{array}$$

The application of the second short method to the total movement of the water molecule also gives the same character representation, as shown below.

$\chi_E = 3 \times 3 = 9$

$\chi_{C_{2(z)}} = 1 \times 1 = -1$

$\chi_{\sigma_{xz}} = 3 \times 1 = 3$

$\chi_{\sigma_{yz}} = 1 \times 1 = 1$

Thus, the representation is 9 –1 3 1. Method II is applicable when the base chosen is the total movement of a molecule.

4.5 CLASSIFICATION OF REPRESENTATIONS

Representations are classified into *reducible* and *irreducible* representations.

The definitions of these representations and the differences between them are highlighted in Table 4.1.

Table 4.1 Reducible and irreducible representations

Reducible Representation	**Irreducible Representation**
If dimensions of representations can be reduced either using the similarity transformation method or the block factoring method, they are referred to as reducible representations.	Irreducible representations are those representations whose dimensions cannot be reduced by any of the mathematical methods like the block factoring or the similarity transformation methods.
They can be expressed as a combination of various irreducible representations that are possible for a group.	They cannot be expressed as combinations of any other representations.
Reducible representations are derived for the product or for the combination of fundamental bases.	Irreducible representations are derived for fundamental bases.

Similarity transformation and block factoring: As mentioned earlier, it is always possible to find a similarity transformation matrix which will transform all the matrices of a group representations into diagonalised matrices with identical blocks or sub matrices along the diagonals of all matrices. Let A,B,C,D be the matrix representations of a C_{2v} group and P be the similarity transformation matrix, so that

$$P^{-1}AP=A'; P^{-1}BP=B'; P^{-1}CP=C'; P^{-1}DP=D'$$

Just as A,B,C,D represent the group, the new set of diagonalised matrices A′,B′,C′,D′ represent the full group. Let the diagonal matrices corresponding to various symmetry elements be,

$$\begin{pmatrix} P & P & 0 & 0 & 0 \\ P & P & 0 & 0 & 0 \\ 0 & 0 & P & 0 & 0 \\ 0 & 0 & 0 & P & P \\ 0 & 0 & 0 & 0 & P \end{pmatrix} \qquad \begin{pmatrix} Q & Q & 0 & 0 & 0 \\ Q & Q & 0 & 0 & 0 \\ 0 & 0 & Q & 0 & 0 \\ 0 & 0 & Q & Q & Q \\ 0 & 0 & 0 & Q & Q \end{pmatrix}$$

$$\begin{pmatrix} R & R & 0 & 0 & 0 \\ R & 0 & 0 & 0 & 0 \\ 0 & 0 & R & 0 & 0 \\ 0 & 0 & 0 & R & R \\ 0 & 0 & 0 & 0 & R \end{pmatrix} \qquad \begin{pmatrix} S & S & 0 & 0 & 0 \\ 0 & S & 0 & 0 & 0 \\ 0 & 0 & S & 0 & 0 \\ 0 & 0 & 0 & S & S \\ 0 & 0 & 0 & S & S \end{pmatrix}$$

Block factoring is a mathematical method to reduce a representation of higher dimension to one having a lower dimension. It can be observed that each of the above matrices has the same structure. That is, all the non-vanishing elements occur within the same square brackets along the diagonal. Since the block structure is identical for every symmetry operation in the representation, the individual blocks themselves can form lower dimensional representations as shown below.

$$\begin{bmatrix} P & P \\ P & P \end{bmatrix} \quad \begin{bmatrix} Q & Q \\ Q & Q \end{bmatrix} \quad \begin{bmatrix} R & R \\ R & R \end{bmatrix} \quad \begin{bmatrix} S & S \\ S & S \end{bmatrix}$$

$$[P] \quad [Q] \quad [R] \quad [S]$$

$$\begin{bmatrix} P & P \\ 0 & P \end{bmatrix} \quad \begin{bmatrix} Q & Q \\ Q & Q \end{bmatrix} \quad \begin{bmatrix} R & R \\ 0 & R \end{bmatrix} \quad \begin{bmatrix} S & S \\ S & S \end{bmatrix}$$

The determination of similarity transformation matrix and the method for diagonalisation of a 2 × 2 matrix is discussed in Appendix I.

However, the dimension of a given reducible representation can be very easily reduced using the Orthogonality Theorem (which is dealt with in the next chapter).

5

Irreducible Representations and Character Tables

5.1 Irreducible Representations (IR)

In Chapter 4, we have learnt about the characteristics of representations, types of representations and how to derive them. If the representations are derived from fundamental bases, they are irreducible representations and their dimensions cannot be reduced further by any mathematical method. In the case of reducible representations, they can be expressed as combinations of irreducible representations of a point group. Irreducible representations reflect the essence of a point group. To learn more about point groups and the basis of representations, it is essential to know all possible irreducible representations of a point group and also their corresponding bases. Therefore, we will now focus on such non-equivalent irreducible representations.

In this chapter we will learn the *Great Orthogonality Theorem* which deals with the relationship between various matrix elements of irreducible representations. This theorem plays a key role in solving complex quantum mechanical problems through irreducible representations.

5.2 The Great Orthogonality Theorem

The theorem is mathematically represented as follows.

$$\sum_R [\Gamma^i_{mn}(R)]^* [\Gamma^j_{m'n'}(R)] = (h/\sqrt{l_i l_j})\ \delta_{ij}\ \delta_{mm'}\ \delta_{nn'}$$

where $\Gamma^i_{mn}(R)$ is a matrix element belonging to the mth row and the nth column of a symmetry operation R in the ith irreducible representation. $\Gamma^j_{m'n'}(R)$ is the matrix element belonging to the m'^{th} row and the n'^{th} column of a symmetry operation R in the jth irreducible representation. l_i and l_j are the dimensions of the ith and the jth irreducible representations and h is the order of the group. It is necessary to take the complex conjugate of one factor on the left side, whenever imaginary or complex numbers are involved.

$$\delta_{ij} = 0 \quad \text{if} \quad i \neq j$$

$$\text{and} \quad \delta_{ij} = 1 \quad \text{when} \quad i = j.$$

Thus, six solutions are simultaneously introduced in the theorem by the three δs. If $i = j$, $m = m'$, $n = n'$, the above equation reduces to

$$\sum_R [\Gamma^i_{mn}(R))]^* \, [\Gamma^i_{m'n'}(R)] = 0 \qquad (5.1)$$

Similarly if $i = j$, $m = m'$, $n \neq n'$, the equation reduces to

$$\sum_R [\Gamma^i_{mn}(R)]^* \, [\Gamma^i_{mn'}(R)] = 0 \qquad (5.2)$$

If $i \neq j$, $m = m'$, $n = n'$, then,

$$\sum_R [\Gamma^i_{mn}(R)]^* \, [\Gamma^j_{mn}(R)] = 0 \qquad (5.3)$$

If $i = j$, $m = m$, $n = n'$, then,

$$\sum_R [\Gamma^i_{mn}(R)]^* \, [\Gamma^i_{mn}(R)] = \sum_R [\Gamma^i_{mn}(R)]^2 = h / l_i \qquad (5.4)$$

Equations 5.1, 5.2, 5.3 and 5.4 are simplified expressions of the orthogonality theorem.

5.3 Significance of the Orthogonality Theorem

Every matrix element of a representation is a vector since it indicates the magnitude and direction of transformation of a component of the base chosen; meaning, the Cartesian coordinates or an orbital. Thus, if we choose a representation of dimension 'l' (where l indicates the presence of l number of rows and columns) it will contain l^2 matrix elements or l^2 vectors for every symmetry element of the group. Combination of corresponding matrix elements (matrix elements belonging to the same row and the same column) chosen from each symmetry element of a group forms a h dimensional vector where h is the order of the group. Such h dimensional vectors collectively represent the transformation of a particular component of the base chosen due to the influence of the symmetrical environment of the group. The nature of such h dimensional vectors is well defined by the orthogonality theorem.

The simplified equations 5.1 and 5.2 of the orthogonality theorem, that is,

$$\sum_R [\Gamma^i_{mn}(R)]^* \, [\Gamma^i_{m'n}(R)] = 0 \qquad \text{if } m \neq m' \qquad \text{and}$$

$$\sum_R [\Gamma^i_{mn}(R)]^* [\Gamma^i_{mn'}(R)] = 0 \qquad \text{if } n \neq n'$$

indicate that such h dimensional vectors existing in the same IR of a group are orthogonal to each other since their vector product is zero.

Similarly, Eq. 5.3

$$\sum_R [\Gamma^i_{mn}(R)]^* [\Gamma^j_{mn}(R)] = 0$$

implies that h dimensional vectors which can be derived from the various IRs that are possible for a group, should be mutually orthogonal to each other, therefore reducing their vector product to zero.

The Eq. 5.4, that is,

$$\sum_R [\Gamma^i_{mn}(R)]^* [\Gamma^i_{mn'}(R)] = h / l_i$$

defines the square of the length of the h dimensional vectors derived from various IR of a group. This should be equal to $\frac{h}{l}$ where h is the order of the group and 'l' is the dimension of the IR.

Thus, representations containing h dimensional vectors which satisfy all the conditions of the orthogonality theorem are irreducible representations and vice versa. This can be illustrated by the following hypothetical example. Let the symmetry elements of a point group be A, B and C.

Representation	A	B	C
I	$\begin{pmatrix} a & b \\ c & d \end{pmatrix}$	$\begin{pmatrix} e & f \\ g & h \end{pmatrix}$	$\begin{pmatrix} i & j \\ k & l \end{pmatrix}$
II	a′	b′	c′

From the two dimensional representation, four 3-dimensional vectors can be derived by the combination of corresponding matrix elements, one chosen from each symmetry element of the group. They are represented as

a e i
b f j
c g k
d h l

These four vectors indicate the transformation of a component of the originally chosen base due to the influence of the symmetry of the point group.

The one dimensional representation gives rise to one 3-dimensional vector a′ b′ c′. If these two representations are irreducible representations or non-equivalent representations, all the five 3-dimensional vectors should be mutually orthogonal to each other and the square of the length of every vector should be $\frac{h}{l}$. When these conditions are not satisfied, the representation becomes a reducible representation. However, the great orthogonality theorem leads to five important consequences which in turn are extensively used in various applications of group theory.

5.4 Consequences of the Orthogonality Theorem

(i) The sum of the squares of the dimensions of all the IRs that are possible for a point group should be equal to the order of the group.

$\sum_i li^2 = h$ where l is the dimension of the ith representation and h is the order of the group.

That is, $l_1^2 + l_2^2 + \ldots l_i^2 = h$

where l_1 = dimension of the first IR, l_2 = dimension of the second IR and so on.

Proof: If l_1 is the dimension of the representation, then l_1^2 matrix elements will exist in the representation. Hence l_1^2 number of h dimensional vectors can be derived from such representations. If l_2 is the dimension of the next representation, l_2^2 matrix elements will be present and hence l_2^2 number of h dimensional vectors can be derived by combining the matrix elements belonging to different rows and columns. Thus, the total number of h dimensional vectors that are possible with various IRs of a group are $l_1^2 + l_2^2 + l_3^2 \ldots l_i^2$.

If all the representations are IRs, then all the above h dimensional vectors should be orthogonal to each other according to the orthogonality theorem.

In mathematics, it has been proved beyond doubt that, a maximum of only h number of h dimensional vectors can be orthogonal with respect to each other. Therefore, the total number of h dimensional vectors which is equal to $l_1^2 + l_2^2 + + l_3^2 \ldots l_i^2$, should be equal to h, which is the order of the group.

$$l_1^2 + l_2^2 + \ldots l_i^2 = h \quad \text{or} \quad \sum l_i^2 = h$$

(ii) The absolute square of the length of any character representation is equal to the order of the group, provided the representation is an irreducible representation. That is, the absolute square of the character corresponding to the irreducible representation summed over the symmetry operations of a point group should be equal to the order of the group.

$$\Sigma \, |\chi_i (R)|^2 = h$$

That is, $|\chi_i (R_1)|^2 + |\chi_i (R_2)|^2 + \text{--------} + |\chi_i (R_i)|^2 = h$

where χ_i (R_1) is the character corresponding to the symmetry element R_1 in the ith IR and so on.

Proof: Consider the orthogonality theorem equation.

$$\sum_R [\Gamma^i{}_{mn} (R)]^* [\Gamma^j{}_{m'n'} (R)] = (h/\sqrt{l_i l_j}) \; \delta_{ij} \; \delta_{mm'} \; \delta_{nn'}$$

When $i = j$, $m = n$ and $n' = m'$, $\delta_{ij} = 1$,

$$\sum_R [\Gamma^i{}_{mm}(R)]^* [\Gamma^i{}_{m'm'}(R)] = (h/l_i)\ \delta_{mm'}$$

(As the δs can have values of 0 or 1, the square terms need not be considered.) Summing up over m and m′ on both sides,

$$\sum_m \sum_{m'} \sum_R [\Gamma^i{}_{mm}(R)]^* [\Gamma^i{}_{m'm'}(R)] = \sum_m \sum_{m'} (h/l_i)\ \delta_{mm'}$$

On rearranging,

$$\sum_R \sum_m [\Gamma^i{}_{mm}(R)]^* \sum_{m'} [\Gamma^j{}_{m'm'}(R)] = \sum_{m'} \sum_m (h/l_i)\ \delta_{mm'}$$

But $\sum_m [\Gamma^i{}_{mm}(R)]$ = Sum of diagonal elements = $\chi_i(R)$

Similarly $\sum_{m'} [\Gamma^i{}_{m'm'}(R)]$ = Sum of diagonal elements = $\chi_i(R)$

m and m′ can have values 1, 2, 3 upto l_i, where l_i is the dimension of the representation. If $m = m' = 1$, $\delta_{mm'} = \delta_{11} = 1$ and if $m = m' = 2$, $\delta_{mm'} = \delta_{22} = 1$.

$$\therefore \sum_{m'=1}^{li} \sum_{m=1}^{li} \delta_{mm'} = 1 + 1 + 1 + \ldots + (l_i \text{ times}) \text{ when } m = m' = l_i$$

Substituting all the values in the above equation, we get

$$\sum_R \chi_i(R)\chi_i(R) = \frac{h}{l} \times l = h$$

That is, $\sum_R \chi_i^2(R) = h$

(iii) The product of any two character representations is zero when the representations are irreducible representations. That is, the character representations derived from the IRS of a group are mutually orthogonal.

That is, $\sum_R \chi_i(R)\chi_j(R) = 0$

Proof: Consider the orthogonality theorem equation,

$$\sum_R [\Gamma^i{}_{mn}(R)]^* [\Gamma^j{}_{m'n'}(R)] = (h/\sqrt{l_i l_j})\ \delta_{ij}\ \delta_{mm'}\ \delta_{nn'}$$

When $n = m$, $n' = m'$

$$\sum_R [\Gamma^i{}_{mm}(R)]^* [\Gamma^j{}_{m'm'}(R)] = (h/\sqrt{l_i l_j}) \times 0 = 0$$

Summing up over m and m'

$$\sum_m \sum_{m'} \sum_R [\Gamma^i{}_{mm}(R)]^* [\Gamma^j{}_{m'm'}(R)] = 0$$

Rearranging,

$$\sum_{R}\sum_{m=1}^{l_i} [\Gamma^{i}_{mm}(R)]^{*} \sum [\Gamma^{j}_{m'm'}(R)] = 0$$

$$\sum_{m=1}^{l_i} \Gamma^{i}_{mm}(R) = \chi_i(R)$$

$$\sum_{m'=1}^{l_i} \Gamma^{i}_{m'm'}(R) = \chi_j(R)$$

$$\therefore \sum_{R} \chi_i(R)\, \chi_j(R) = 0$$

(iv) The characters corresponding to the conjugate elements are the same in any IR. Since the transformation matrices of conjugate elements are the matrix and its transpose, their character remains the same. For example, consider the C_{3v} point group. $C_3^{\,1}$ and $C_3^{\,2}$ are conjugate elements.

$$\chi_{C_3^{\,1}} = 2 \cos 120^{\circ} + 1 \; (\because \chi_{C_n} = 2 \cos\theta + 1)$$

$$= 2 \times -\frac{1}{2} + 1 = 0$$

$$\chi_{C_3^{\,2}} = (2 \times \cos 240^{\circ}) + 1 = 2 \times -\frac{1}{2} + 1 = 0$$

(v) The total number of IRs that are possible for a point group is equal to the total number of classes present in a point group. For example, the number of classes for the C_2 point group is 2. That is, C_2 and E. Hence, the number of of IRs = 2. Similarly, for the C_{3v} point group, the number of classes is 3; E, $2C_3$ and $3\,\sigma_v$. Hence, the number of IRs = 3

D_{2h} : 8 classes, 8 IRs
C_s : 2 classes, 2 IRs

Proof: Let the order of the group be h and number of classes be k, where k is less than h. As characters corresponding to conjugate elements remain the same, the elements belonging to the same class will have the same character. Hence, the equation

$$\sum_{R=1}^{h} \chi_i(R)\, \chi_j(R) = 0$$

can be mathematically simplified as

$$\sum_{k=1}^{p} g_p\, \chi_i(R_p)\, \chi_j(R_p) = 0$$

where, g_p is the number of elements belonging to p^{th} class and R_p is one of the symmetry elements belonging to the pth class. Since the product of the two character representations is zero, it indicates that the two representations are orthogonal to each other.

These character representations remain k dimensional in nature. In mathematics, it has been proved that a maximum of k number of k dimensional vectors can be mutually orthogonal to each other. This automatically requires the presence of only k number of character representations which can be orthogonal to each other. Since every IR can give rise to a maximum of one character representation, only k number of IRs are expected which can give rise to k number of mutually orthogonal character representations for the group. Thus, the total number of IRs that are possible for a point group remains equal to the total number of classes present in a point group.

5.5 APPLICATIONS OF THE ORTHOGONALITY THEOREM

(i) Character tables are constructed using the orthogonality theorem.
(ii) Any given reducible representation (RR) can be expressed as a combination of the various IRs possible for a point group.

5.6 CHARACTER TABLES

A character table contains four main areas and gives information about the following.

(i) The name of the point group with the possible symmetry elements classified under appropriate classes.
(ii) The number of IRs possible for a point group.
(iii) The character corresponding to various classes of symmetry elements for all the IRs that are possible for a point group.
(iv) The symmetry and corresponding fundamental bases for every IR.
(v) The respective combined bases for the various IRs of a point group.

For example, the character table for a C_{2V} point group is given below.

Area II			**Area I**		**Area III**	**Area IV**
C_{2V}	E	$C_{2(z)}$	σ_{xz}	σ_{yz}		
A_1	1	1	1	1	Z	x^2, y^2, z^2
A_2	1	1	–1	–1	$R_{(z)}$	xy
B_1	1	–1	–1	1	y, R_x	yz
B_2	1	–1	1	–1	x, R_y	xz

From the character table, we can see that Area I contains the list of symmetry elements possible for a point group with appropriate classes. It also gives the respective characters of each IR possible for a point group and is mainly derived from the orthogonality theorem. Area II contains the Mulliken symbols which indicate the symmetry of the various IRs that are possible for a point group. In Area III, the corresponding fundamental bases for the various IRs are listed. Area IV contains the combined bases for the various IRs.

5.6.1 Construction of Character Tables

Area I of the character table can be derived using the orthogonality theorem as given below.

(i) The total number of IRs of a group is equal to the total number of classes present in the group in accordance with the axiom derived from the orthogonality theorem which states that the number of IRs is equal to the number of classes in the group.

(ii) The dimensions of the various IRs can be derived using the following axiom derived from the orthogonality theorem,

$$\therefore \sum_i l_i^2 = h$$

(iii) Using other axioms derived from the orthogonality theorem,

$$\sum_R \chi_i^2(R) = h \quad \text{and} \quad \sum_R \chi_i(R)\, \chi_j(R) = 0,$$

it is possible to derive characters corresponding to every symmetry element of the group.

Area II of the character table can be derived based on following conventions reported by Mulliken.

(i) One-dimensional representations are represented by A or B, two-dimensional representations by E and three dimensional representations by T.

(ii) One-dimensional representations which are symmetric to the principal axis are represented by A and those which are antisymmetric to the principal axis are represented by B. (Symmetric: the character is positive; antisymmetric: the character is negative.)

(iii) If the representations are symmetric to *i*, a subscript '*g*' is added; whereas, if the representations are antisymmetric to *i*, a subscript *u* is added.

(iv) A subscript '1' is added to representations which are symmetric with respect to σ_v and '2' is added to those representations which are antisymmetric with respect to σ_v.

(v) Single prime (′) is added as a superscript for those representations which are symmetric to σ_h and double prime (″) is added as a superscript for those operations which are antisymmetric to σ_h.

Area III of the character table is derived by determining the influence of the symmetry operations of a group on the various fundamental bases. Generally $\vec{x}, \vec{y}, \vec{z}$ and rotation about the x, y and z axes (R_x, R_y, R_z) are considered as fundamental bases.

Area IV of the character table consisting of product vectors is derived using the direct product concept which will be dealt with in Chapter 6.

5.6.2 Character Table for the C_{2V} Point Group

(i) The total number of IRs possible for a C_{2V} point group is 4 since the point group contains 4 classes of symmetry operations.

(ii) Let the dimensions of the four IRs be l_1, l_2, l_3 and l_4. The order of the group is 4.

$\therefore\ l_1^2 + l_2^2 + l_3^2 + l_4^2 = 4.$

As the dimensions cannot be zero,

$l_1^2 = l_2^2 = l_3^2 = l_4^2 = 1.$

Therefore, all the four representations are one-dimensional representations.

(iii) Since the character corresponding to the identity element is the dimension of the representation, in all the four IRs, $\chi_E = 1$. The characters corresponding to $C_{2(z)}$, σ_{xz} and σ_{yz} will be (x_1, y_1, z_1), (z_2, y_2, z_2), (x_3, y_3, z_3) and (x_4, y_4, z_4) respectively for the four IRs.

(iv) The square of the character representation is equal to the order of the group.

$\therefore\ 1^2 + x_i^2 + y_i^2 + z_i^2 = 4.$

where, i = 1, 2, 3 or 4 corresponding to the four IRs.
As x_i, y_i and z_i cannot exceed the value 1,

$x_i^2 = y_i^2 = z_i^2 = 1.$

$\therefore\ x_i = \pm 1, y_i = \pm 1$ and $z_i = \pm 1.$

(v) To decide the sign of the characters (x_i, y_i, z_i), the expression $\sum_R \chi_i(R)\ \chi_j(R)$ is used.

Substituting the characters from four different IRs of the group, we get

$$1 \times 1 + (x_1 \times x_2) + (y_1 \times y_2) + (z_1 \times z_2) = 0$$

$$1 \times 1 + (x_1 \times x_3) + (y_1 \times y_3) + (z_1 \times z_3) = 0$$

$$1 \times 1 + (x_1 \times x_4) + (y_1 \times y_4) + (z_1 \times z_4) = 0$$

All these equations will be simultaneously satisfied when all the characters are positive in the first IR, and in the remaining IRs, the characters of two symmetry elements are positive, and the characters of two IRs are negative.

Therefore, Area I of the C_{2V} character table is

	E	$C_{2(Z)}$	σ_{xz}	σ_{yz}
I	1	1	1	1
II	1	1	–1	–1
III	1	–1	–1	1
IV	1	–1	1	–1

The highlights of Area II are,

(i) Among the four one dimensional IRs, two are symmetric (+ve character) with respect to the principal C_2 axis and are hence labelled as A. The remaining two IRs that are antisymmetric with respect to the principal C_2 axis (–ve character) are labelled as B representations.

(ii) Representations that are symmetric to σ_v (σ_{yz}) are indicated as A_1 and B_1 whereas representations that are antisymmetric to σ_v are labelled as A_2 and B_2 respectively.

The highlights of Area III are,

The symmetry of the fundamental bases $[\vec{x}, \vec{y}, \vec{z}, R_x, R_y, R_z]$ can be determined by performing the symmetry operations of a group on them and arriving at the corresponding characters using the following short method.

$$\chi_R = \text{Number of bases which do not change position and direction} - \text{Number of bases which do not change position but reverse the direction}$$

Choosing $\vec{z}$ as base,

$$\chi_E = 1 - 0 = 1$$
$$\chi_{C_{2(z)}} = 1 - 0 = 1$$
$$\chi_{\sigma_{xz}} = 1 - 0 = 1$$
$$\chi_{\sigma_{yz}} = 1 - 0 = 1$$

$\therefore$ $\vec{z}$ has A_1 symmetry. Similarly it can be shown that $\vec{y}$ has B_1 symmetry and $\vec{x}$ has B_2 symmetry. Choosing R_x as base, $[R_x = -ih\,(y\frac{\partial}{\partial z} - z\frac{\partial}{\partial y})]$

$$\chi_E = 1$$
$$\chi_{C_{2(z)}} = -1$$
$$\chi_{\sigma_{xz}} = -1$$
$$\chi_{\sigma_{yz}} = 1$$

R_x has B_1 symmetry. Similarly, R_y has B_2 symmetry and R_z has A_2 symmetry.

Note: The character table of the C_{2h} group is similar to that of the C_{2V} group except for the Mulliken's symbols.

5.6.3 Character Table for the C_{3V} Point Group

(i) The total number of IRs possible for the C_{3V} point group is three since this point group contains three classes of symmetry operations.

(ii) Let the dimensions of three the IRs be l_1, l_2 and l_3, and the order of the group be six.

$\therefore\ l_1^2 + l_2^2 + l_3^2 = 6.$

As the dimensions cannot be zero, the only possible solution is $l_1^2 = l_2^2 = 1$ and $l_3^2 = 4$.

That is, $l_1 = l_2 = 1$ and $l_3 = 2$.

Thus, among the three IRs, two are one-dimensional representations and one is a two-dimensional representation.

(iii) The character corresponding to the identity element in any representation is the dimension of that representation.

$\therefore\ \chi_E = 1$ for one-dimensional representations and $\chi_E = 2$ for two-dimensional representations. The characters corresponding to the C_3 and σ_v classes of the C_{3V} group will be (x_1, y_1), (x_2, y_2) and (x_3, y_3) respectively in the three IRs.

(iv) Since $\sum_{p=1}^{k} g_p\, \chi_i\,(R)^2 = h$ for a one-dimensional representation of C_{3V} point group, $1^2 + 2\,x_i^2 + 3\,y_i^2 = 6$ where ($i = 1$ and 2). x_i and y_i can not exceed 1.

$x_i^2 = 1$ and $y_i^2 = 1$

$\therefore\ x_i = \pm 1$ and $y_i = \pm 1$.

For the two-dimensional representation of the group,

$$2^2 + 2x_3^2 + 3y_3^2 = 6$$

The possible solutions are, $x_3^2 = 1$ and $y_3^2 = 0$

$\therefore\ x_3 = \pm 1$ and $y_3 = 0$.

(v) To decide about the sign of character, equation $\sum_{p=1}^{k} g_p\, \chi_i\,(R)\, \chi_j\,(R) = 0$ is used.

Substituting the characters of the one-dimensional representations in the above equation we get,

$(1\times1) + (2x_1 \times x_2) + (3y_1 \times y_2) = 0$

$\therefore\ x_1 = x_2 = +1,\ y_1 = +1$ and $y_2 = -1$.

Similarly, by considering a one-dimensional representation and a two-dimensional representation, we have

$(1\times2) + (2x_1 \times x_3) + (3y_1 \times y_3) = 0$

$2 + 2x_1\, x_3 + 3y_1\, y_3 = 0$

As $x_1 = +1$, $y_1 = +1$ and $y_3 = 0$,

$x_3 = -1$.

Hence Area I of the character table is,

	E	$2C_3$	$3\sigma_v$
I	1	1	1
II	1	1	–1
III	2	–1	0

The highlights of Area II are,

(i) The one-dimensional IR symmetric to $C_{3(Z)}$ and σ_v is labelled as A_1 and the other one symmetric to $C_{3(Z)}$ and antisymmetric to σ_v is indicated as A_2.

(ii) The two-dimensional representation is labelled as E.

The highlights of Area III are,

The symmetry of the fundamental bases $[\vec{X}, \vec{Y}, \vec{Z}$, and $R_x, R_y, R_z]$ can be determined by performing symmetry operations of a group and then arriving at the corresponding characters using a short method. Choosing $\vec{Z}$ as the base,

$$\chi_E = 1 - 0 = 1$$
$$\chi_{C_{3(z)}} = 1 - 0 = 1$$
$$\chi_{\sigma_v} = 1 - 0 = 1$$

$\therefore \vec{Z}$ has A_1 symmetry.

The X and Y axes are inseparable in the C_{3V} point group due to the C_3 operation (since the new position of $\vec{X}$ and $\vec{Y}$ can be expressed only by the combination of original $\vec{X}$ and $\vec{Y}$).

Thus, choosing $\vec{X}$ and $\vec{Y}$ together as the base,

$$\chi_E = 2 - 0 = 2$$
$$\chi_{C_{3(z)}} = 2 \cos 120° + 1 = -1$$
$$\chi_{\sigma_v} = 1 - 1 = 0$$

$\therefore \vec{X}$ and $\vec{Y}$ together have '*E*' symmetry. Similarly R_x and R_y together have '*E*' symmetry and R_z has A_2 symmetry.

5.7 REDUCTION OF A REDUCIBLE REPRESENTATION

Any given reducible representation can be expressed as a combination of the various IRs of a point group using the orthogonality theorem with the help of the character table.

Consider a point group consisting of a number of IRs represented by $a_1, a_2, a_3, \ldots\ldots\ldots a_i$.

Let the reducible representation $= n_1 a_1 + n_2 a_2 + \ldots\ldots\ldots\ldots\ldots + n_i a_i$ (5.5) where n_1, n_2 .... are the number of times the corresponding IRs are present in the reducible representation. Therefore, the character corresponding to a symmetry operation *R* in a RR can be represented by a combination of characters corresponding to the same symmetry operation R in various IRs.

That is, $\chi_{R(RR)} = n_1 \chi_{1(R)} + n_2 \chi_{2(R)} + \ldots\ldots\ldots\ldots + n_i \chi_{i(R)}$ (5.6)

where $\chi_{R(RR)}$ represents the character corresponding to the symmetry operation *R* in a reducible representation. That is, $\chi_{1(R)}$ represents the character corresponding to the symmetry operation *R* in the first IR, $\chi_{2(R)}$ represents the character corresponding to the symmetry operation R in the second IR and so on. Multiplying Eq. 5.6 by $\chi_{1(R)}$ and summing up over all symmetry operations.

$$\sum_R \chi_R^*(RR).\chi_1(R) = n_1 \sum_R \chi_1^*(R).\chi_1(R) + n_2 \sum_R \chi_2^*(R).\chi_1(R) + \ldots\ldots\ldots + n_i \sum_R \chi_i^*(R).\chi_1(R) \quad (5.7)$$

When the consequence of the orthogonality theorem

$\sum_R \chi_i(R)\,\chi_j(R) = 0$ is applied to Eq. 5.7, all terms except the first on the R.H.S get reduced to zero.

$$\therefore \sum_R \chi_R^*(RR).\chi_1(R) = n_1 \sum_R \chi_1(R)^2 \quad (5.8)$$

But, $\sum_R \chi_1(R)^2 = h$ according to the orthogonality theorem.

$$\therefore \sum_R \chi_R^*(RR).\chi_1(R) = n_1 h \quad (5.9)$$

and $$n_1 = \frac{1}{h} \sum_R \chi_R(RR).\chi_1(R) \quad (5.10)$$

Similarly, n_2 can be found as follows

$$\sum_R \chi_R(RR).\chi_2(R) = n_2 \sum_R \chi_2(R)^2 = n_2 h$$

$$n_2 = \frac{1}{h} \sum_R \chi_R(RR).\chi_2(R)$$

$$\therefore n_i = \frac{1}{h} \sum_R \chi_R(RR).\chi_i(R)$$

Thus, $n_1, n_2, \ldots, n_i$ can be determined easily using the orthogonality theorem.

Example 5.1

A RR of the C_{2V} point group, 3 1 –1 1 can be reduced using the orthogonality theorem and the character table.

	E	$C_{2(Z)}$	σ_{xz}	σ_{yz}
A_1	1	1	1	1
A_2	1	1	–1	–1
B_1	1	–1	–1	1
B_2	1	–1	1	–1

$$\text{No. of } A_1 = \frac{1}{h}\sum_R \chi_{R(RR)} \cdot \chi_{R(A_1)}$$

$$= \frac{1}{h}[\chi_{E(RR)} \cdot \chi_{E(A_1)} + \chi_{C_{2z}(RR)} \cdot \chi_{C_{2z}(A_1)} + \chi_{\sigma_{xz}(RR)} \cdot \chi_{\sigma_{xz}(A_1)} + \chi_{\sigma_{yz}(RR)} \cdot \chi_{\sigma_{yz}(A_1)}]$$

$$= \frac{1}{4}[(3\times1) + (1\times1) + (-1\times1) + (1\times1)]$$

$$= \frac{4}{4} = 1.$$

$$\text{No. of } A_2 = \frac{1}{h}\sum_R \chi_{R(RR)} \cdot \chi_{R(A_2)}$$

$$= \frac{1}{h}[\chi_{E(RR)} \cdot \chi_{E(A_2)} + \chi_{C_{2z}(RR)} \cdot \chi_{C_{2z}(A_2)} + \chi_{\sigma_{xz}(RR)} \cdot \chi_{\sigma_{xz}(A_2)} + \chi_{\sigma_{yz}(RR)} \cdot \chi_{\sigma_{yz}(A_2)}]$$

$$= \frac{1}{4}[(3\times1) + (1\times1) + (-1\times-1) + (1\times-1)]$$

$$= \frac{4}{4} = 1.$$

$$\text{No. of } B_1 = \frac{1}{h}\sum_R \chi_{R(RR)} \cdot \chi_{R(B_1)}$$

$$= \frac{1}{h}[\chi_{E(RR)} \cdot \chi_{E(B_1)} + \chi_{C_{2z}(RR)} \cdot \chi_{C_{2z}(B_1)} + \chi_{\sigma_{xz}(RR)} \cdot \chi_{\sigma_{xz}(B_1)} + \chi_{\sigma_{yz}(RR)} \cdot \chi_{\sigma_{yz}(B_1)}]$$

$$= \frac{1}{4}[(3\times1) + (1\times1) + (-1\times-1) + (1\times-1)]$$

$$= \frac{4}{4} = 1.$$

$$\text{No. of } B_2 = \frac{1}{h}\sum_R \chi_{R(RR)} \cdot \chi_{R(B_2)}$$

$$= \frac{1}{h}[\chi_{E(RR)} \cdot \chi_{E(B_2)} + \chi_{C_{2z}(RR)} \cdot \chi_{C_{2z}(B_2)} + \chi_{\sigma_{xz}(RR)} \cdot \chi_{\sigma_{xz}(B_2)} + \chi_{\sigma_{yz}(RR)} \cdot \chi_{\sigma_{yz}(B_2)}]$$

$$= \frac{1}{4}[(3\times1) + (1\times-1) + (-1\times1) + (1\times-1)]$$

$$= 0.$$

$\therefore\ R = A_1 + A_2 + B_1$

Example 5.2

To find the RR for the C_{3V} point group that has less number of classes than the order.

A RR of the C_{3V} point group 4 1 0 can be expressed in terms of IRs of the point group using the orthogonality theorem. When 'k' number of classes and 'h' number of symmetry elements are present,

$$n_i = \frac{1}{h}\sum^{k} g_p\, \chi_{R}(RR).\ \chi_i(R)$$

The C_{3V} character table is given below.

C_{3V}	**E**	$2C_3$	$3\sigma_v$
A_1	1	1	1
A_2	1	1	–1
E	2	–1	0

$$\text{No. of } A_1 = \frac{1}{6}[(1\times4\times1) + (2\times1\times1) + (3\times0\times1)] = \frac{6}{6} = 1$$

$$\text{No. of } A_2 = \frac{1}{6}[(1\times4\times1) + (2\times1\times1) + (3\times0\times1)] = \frac{6}{6} = 1$$

$$\text{No. of E} = \frac{1}{6}[(1\times4\times2) + (2\times1\times-1) + 0] = \frac{6}{6} = 1$$

$\therefore\ R = A_1 + A_2 + E$

This chapter thus highlights the Orthogonality Theorem and how it is useful in deriving other important theorems of irreducible representations and construction of character tables. The important role of the Orthogonality Theorem in expressing a given RR as a combination of IR of a group paves the way for the applications of group theory.

6

Symmetry of Hybrid Orbitals

6.1 Determination of Symmetry of Hybrid Orbitals – General Discussion

The central atom of any molecule with more than two atoms can be considered to have undergone hybridisation. This concept is invoked to explain the strong nature of the bonds and their directional nature. In such molecules, the hybrid orbitals are used as the base for the representation to determine the symmetry of the hybridised orbitals. All the symmetry operations of the point group of the molecule are listed with appropriate classes. These symmetry operations are performed on the hybrid orbitals and their respective transformation matrices can be derived as described in Chapter 5. A collection of characters of such transformation matrices of various classes of symmetry operations of the group can generate a character representation corresponding to the hybrid orbitals.

However, such a representation can also be derived by using the following simple method.

$\chi_{R(\text{hybrid orbital})}$ = Number of orbitals which do not change the position and direction – Number of orbitals which do not change the position but simply reverse direction when a symmetry operation is performed.

The RR corresponding to the hybrid orbitals can be expressed as a combination of the various IRs of the point group of the molecule using the great orthogonality theorem and the character table. The IRs corresponding to the hybrid orbitals indicate their symmetry.

By referring to Area III and Area IV of the character table, the fundamental bases and the combined bases corresponding to the IRs present in the RR of the hybrid orbitals can be derived. This in turn gives us an idea about the nature of hybridisation.

6.2 SYMMETRY OF HYBRIDISATION OF VARIOUS MOLECULES

6.2.1 Methane

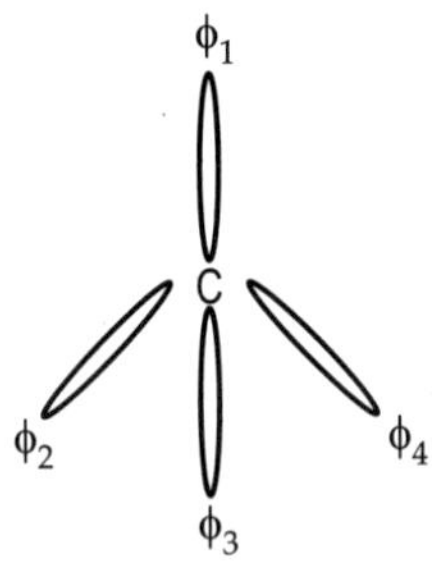

Let the four hybrid orbitals of methane be represented as ϕ_1, ϕ_2, ϕ_3 and ϕ_4. These are taken as bases for which the representation has to be derived under an environment that is tetrahedrally symmetrical. The various classes of symmetry operations in a tetrahedral point group include E, C_3, C_2, S_4 and σ_d.

The character representation of such hybrid orbitals is obtained by performing symmetry operations on hybrid orbitals and deriving their respective transformation matrices, or by using the simple method described earlier. By using the short method,

$$\chi_{E(\text{hybrid orbital})} = 4 - 0 = 4$$

$$\chi_{C_3\,(\text{hybrid orbital})} = 1 - 0 = 1$$

$$\chi_{C_2\,(\text{hybrid orbital})} = 0 - 0 = 0$$

$$\chi_{S_4\,(\text{hybrid orbital})} = 0 - 0 = 0$$

$$\chi_{\sigma_d\,(\text{hybrid orbital})} = 2 - 0 = 2$$

∴ The character representation is

E	$8C_3$	$3C_2$	$6S_4$	$6\sigma_d$
4	1	0	0	2

Using the orthogonality theorem and the character table, this can be reduced to give the IRs of the group as given below.

Td	E	$8C_3$	$3C_2$	$6S_4$	$6\sigma_d$
$\chi_{R(\text{Hybrid orbital})}$	4	1	0	0	2
A_1	1	1	1	1	1
A_2	1	1	1	–1	–1
E	2	–1	2	0	0
T_1	3	0	–1	1	–1
T_2	3	0	–1	–1	1

$$\text{No. of } A_1 = 1/24\ [(1 \times 4 \times 1) + (8 \times 1 \times 1) + (3 \times 0 \times 1) + (6 \times 0 \times 1) + (6 \times 2 \times 1)]$$

$$= \frac{1}{24}(4 + 8 + 12) = 1$$

No. of $A_2 = 1/24\,[\,(1 \times 4 \times 1) + (8 \times 1 \times 1) + 0 + 0 + (6 \times 2 \times -1)\,]$

$$= \frac{1}{24}(12 - 12) = 0$$

No. of $E = 1/24\,[(1 \times 4 \times 2) + (8 \times 1 \times -1) + 0 + 0 + (6 \times 2 \times 0)]$

$$= \frac{1}{24}(8 - 8) = 0$$

No. of $T_1 = 1/24\,[\,(1 \times 4 \times 3) + 0 + 0 + 0 + (6 \times 2 \times -1)\,]$

$$= \frac{1}{24}(12 - 12) = 0$$

No. of $T_2 = 1/24\,[\,(1 \times 4 \times 3) + 0 + 0 + 0 + (6 \times 2 \times 1)\,]$

$$= \frac{1}{24}(12 + 12) = 1$$

RR (hybrid orbitals) = $A_1 + T_2$

A_1 is the most symmetric representation and it must therefore correspond to the highly symmetric *s* orbital. Area III of the character table indicates the x, y, z vectors corresponding to the T_2 symmetry which represent the p_x, p_y and the p_z orbitals. Thus, we can say that the hybrid orbitals are formed by the mixing of one *s* orbital and three *p* orbitals resulting in sp^3 hybridisation.

Area IV indicates xy, yz, xz for T_2 symmetry in the character table. They represent orbitals that are projected along the xy, yz and xz planes. The corresponding orbitals are the d_{xy}, d_{yz}, and d_{xz}. So, it is possible for the hybridisation to be sd^3 as well. But, in the carbon atom, the excitation of electrons from the *2p* level to the *3d* level involves higher energy and hence sd^3 hybridisation is energetically not favourable even though it is symmetrically allowed. Thus, in methane, sp^3 hybridisation is symmetrically and energetically favoured.

6.2.2 BCl_3

BCl_3 belongs to the D_{3h} point group and the various classes of symmetry elements are E, C_3, C_2, σ_h, S_3 and σ_v. Character representation of the hybrid orbitals is obtained by using the short method as follows.

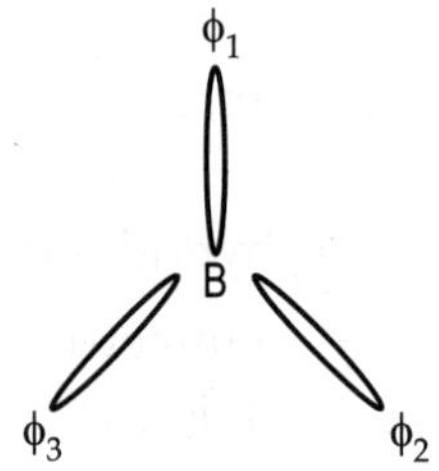

$$\chi_E = 3 - 0 = 3$$
$$\chi_{C_3} = 0 - 0 = 0$$
$$\chi_{C_2} = 1 - 0 = 1$$

$$\chi_{\sigma_h} = 3 - 0 = 3$$
$$\chi_{S_3} = 0 - 0 = 0$$
$$\chi_{\sigma_v} = 1 - 0 = 1$$

The character representation is

E	$2C_3$	$3C_2$	σ_h	$2S_3$	$3\sigma_v$
3	0	1	3	0	1

By using the orthogonality theorem and the character table, this can be reduced to the IRs of the group as follows.

D_{3h}	E	$2C_3$	$3C_2$	σ_h	$2S_3$	$3\sigma_v$	
χ_R (hybrid orbitals)	3	0	1	3	0	1	
A_1'	1	1	1	1	1	1	
A_2'	1	1	–1	1	1	–1	
E'	2	–1	0	2	–1	0	x, y, x^2-y^2, xy
A_1''	1	1	1	–1	–1	–1	
A_2''	1	1	–1	–1	–1	1	
E''	2	–1	0	–2	1	0	

No. of $A_1' = 1/12\,[(1 \times 3 \times 1) + (2 \times 0 \times 1) + (3 \times 1 \times 1) + (1 \times 3 \times 1) + (2 \times 0 \times 1) + (3 \times 1 \times 1)]$

$$= \frac{1}{12}(3 + 0 + 3 + 3 + 0 + 3) = 1$$

No. of $A_2' = \frac{1}{12}(3 + 0 - 3 + 3 + 0 - 3) = 0$

No. of $E' = \frac{1}{12}(6 + 0 + 0 + 6 + 0 + 0) = 1$

No. of $A_1'' = \frac{1}{12}(3 + 0 + 3 - 3 + 0 - 3) = 0$

No. of $A_2'' = \frac{1}{12}(3 + 0 - 3 - 3 + 0 + 3) = 0$

No. of $E'' = \frac{1}{12}(6 + 0 + 0 - 6 + 0 + 0) = 0$

$\therefore$ RR (hybrid orbitals) = $A_1' + E'$

A_1' is symmetric to all the symmetry elements and therefore, it represents the *s* orbital. E' corresponds to (x, y) in Area III and ($x^2 - y^2$, xy) in area IV of the character table.

∴ The corresponding orbitals should be p_x and p_y or d_{xy} and $d_{x^2-y^2}$. Therefore, the hybridisation may be either sp^2 or sd^2. But, sd^2 is not energetically favourable even though it is symmetrically allowed. This is because, it requires the excitation of electrons from the $2p$ to $3d$ level which involves higher energy.

6.2.3 XeF_4

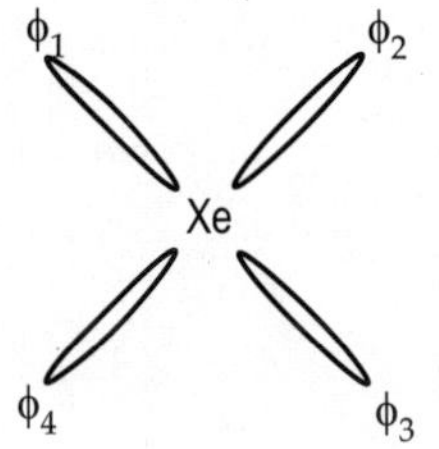

XeF_4 belongs to the D_{4h} point group. Let ϕ_1, ϕ_2, ϕ_3 and ϕ_4 be the four hybrid orbitals of XeF_4 chosen as the base for which the representation has to be derived for a square planar environment. Various classes of symmetry elements of the D_{4h} point group are E, $2C_4$, $C_4^2 = C_2$, $2C_2'$, $2C_2''$, i, $2S_4$, σ_h, σ_v and σ_d.

The character representation of the hybrid orbitals is obtained by using the above mentioned simple method.

$$\chi_E = 4 - 0 = 4 \qquad \chi_i = 0 - 0 = 0$$

$$\chi_{C_4} = 0 - 0 = 0 \qquad \chi_{S_4} = 0 - 0 = 0$$

$$\chi_{C_2} = 0 - 0 = 0 \qquad \chi_{\sigma_h} = 4 - 0 = 4$$

$$\chi_{C'_2} = 0 - 0 = 0 \qquad \chi_{\sigma_v} = 0 - 0 = 0$$

$$\chi_{C''_2} = 2 - 0 = 2 \qquad \chi_{\sigma_d} = 2 - 0 = 2$$

∴ The character representation is

E	$2C_4$	$C_4^2 = C_2$	$2C_2'$	$2C_2''$	i	$2S_4$	σ_h	$2\sigma_v$	$2\sigma_d$
4	0	0	0	2	0	0	4	0	2

By using the orthogonality theorem and the character table, this can be reduced to generate the IRs of the group as follows.

D_{4h}	E	$2C_4$	$C_4^2 = C_2$	$2C_2'$	$2C_2''$	i	$2S_4$	σ_h	$2\sigma_v$	$2\sigma_d$	
χ_R (hybrid orbital)	4	0	0	0	2	0	0	4	0	2	
A_{1g}	1	1	1	1	1	1	1	1	1	1	x^2+y^2; z^2
A_{2g}	1	1	1	–1	–1	1	1	1	–1	–1	Rz
B_{1g}	1	–1	1	+1	–1	+1	–1	1	1	–1	x^2-y^2
B_{2g}	1	–1	1	–1	1	1	–1	1	+1	1	xy
E_g	2	0	–2	0	0	2	0	–2	0	0	(Rx, Ry) (xz, yz)
A_{1u}	1	1	1	1	1	–1	–1	–1	–1	–1	
A_{2u}	1	1	1	–1	–1	–1	–1	–1	1	1	Tz
B_{1u}	1	–1	1	1	–1	–1	1	–1	–1	1	
B_{2u}	1	–1	1	–1	1	–1	1	–1	1	–1	
E_u	z	0	–2	0	0	–2	0	2	0	0	(Tx, Ty)

$$\text{No. of } A_{1g} = \frac{1}{16}\,[(1\times4\times1)+(2\times0\times1)+(1\times0\times1)+(2\times0\times1)+(2\times2\times1)+$$
$$(1\times0\times1)+(2\times0\times1)+(1\times4\times1)+(2\times0\times1)+(2\times2\times1)]$$
$$= \frac{1}{16}\,(4+0+0+0+4+0+0+4+0+4) = 1$$

$$\text{No. of } A_{2g} = \frac{1}{16}\,(4+0+0-0-4+0+0+4-0-4) = 0$$

$$\text{No. of } B_{1g} = \frac{1}{16}\,(4-0+0-0+4-0+0+4+0-4) = 0$$

$$\text{No. of } B_{2g} = \frac{1}{16}\,(4-0+0-0+4+0-0+4-0+4) = 1$$

$$\text{No. of } A_{1u} = \frac{1}{16}\,(4+0+0+0+4+0+0-4+0-4) = 0$$

$$\text{No. of } A_{2u} = \frac{1}{16}\,(4+0+0+0-4+0+0-4+0+4) = 0$$

$$\text{No. of } B_{1u} = \frac{1}{16}\,(4+0+0+0-4+0+0-4+0+4) = 0$$

$$\text{No. of } B_{2u} = \frac{1}{16}\,(4+0+0+0-4+0+0+4+0-4) = 0$$

$$\text{No. of } E_{u} = \frac{1}{16}\,(8+0-0+0+0-0+0+8+0+0) = 1$$

$\therefore$ RR (hybrid orbitals) = $A_{1g} + B_{2g} + E_u$

A_{1g} is symmetric to all symmetry elements and it represents the *s* orbital. B_{2g} corresponds to xy (Area IV of the character table) and it represents a d_{xy} orbital. E_u corresponds to (x, y) and it represents the p_x and p_y orbitals. Thus, the hybridisation involved in XeF_4 is dsp^2.

6.2.4 H_2O

H_2O belongs to the C_{2V} point group and it has a distorted tetrahedron geometry as it has two bonded and two lone pairs of electrons. Let ϕ_1, ϕ_2, ϕ_3 and ϕ_4 be the four hybrid orbitals taken as the base. The character representation of these orbitals can be derived under a tetrahedral environment using the short method as follows.

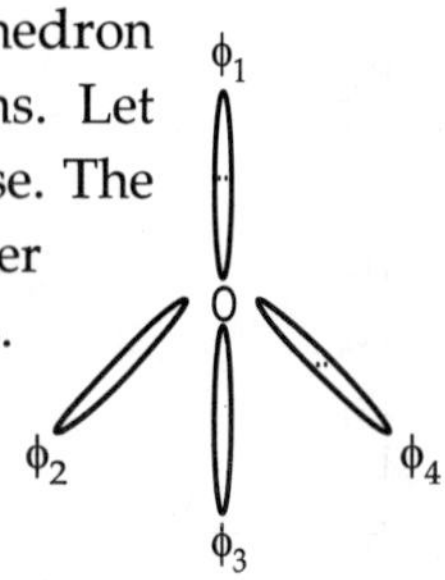

$$\chi_E = 4 - 0 = 4$$

$$\chi_{C_2} = 0 - 0 = 0$$

$$\chi_{\sigma_{xz}} = 2 - 0 = 2$$

$$\chi_{\sigma_{yz}} = 2 - 0 = 2$$

The character representation of hybrid orbitals is

E	$C_{2(z)}$	σ_{xz}	σ_{yz}
4	0	2	2

By using the orthogonality theorem and the character table, it can be reduced to generate the IRs of the group as follows.

C_{2V}	E	$C_{2(z)}$	σ_{xz}	σ_{yz}		
χ_R (hybrid orbitals)	4	0	2	2		
A_1	1	1	1	1		x^2, y^2, z^2
A_2	1	1	–1	–1	R_z	xy
B_1	1	–1	1	–1	x, R_y	xz
B_2	1	–1	–1	1	y, R_x	yz

$$\text{No. of } A_1 = 1/4\,[(1 \times 4 \times 1) + (1 \times 0 \times 1) + (1 \times 2 \times 1) + (1 \times 2 \times 1)]$$

$$= \frac{1}{4}[8] = 2$$

$$\text{No. of } A_2 = \frac{1}{4}[4 + 0 - 2 - 2] = 0$$

$$\text{No. of } B_1 = \frac{1}{4}[4 + 0 + 2 - 2] = 1$$

$$\text{No. of } B_2 = \frac{1}{4}[4 + 0 - 2 + 2] = 1$$

$\therefore$ RR (hybrid orbitals) $= 2A_1 + B_1 + B_2$

The orbitals corresponding to the symmetry A_1, B_1 and B_2 combine to give hybrid orbitals. From the Areas III and IV of the character table, it can be seen that these symmetries correspond to the s, p_z, p_y and p_x orbitals. Thus, the hybridisation involved in H_2O is sp^3.

6.2.5 NH_3

The ammonia molecule belongs to the C_{3V} point group. In ammonia, there are three bonded and one lone-pair of electrons. Let the four hybrid orbitals ϕ_1, ϕ_2, ϕ_3 and ϕ_4 be chosen as the base for arriving at the character representation. The character representation of the hybrid orbitals can be derived as follows.

$$\chi_E = 4 - 0 = 4$$

$$\chi_{C_3} = 1 - 0 = 1$$

$$\chi_{\sigma_v} = 2 - 0 = 2$$

Thus, the character representation is

E	C_3	σ_v
4	1	2

By using the orthogonality theorem and the character table, it can be reduced to give the IRs of the group as follows.
The C_{3V} character table is shown below.

C_{3V}	E	$2C_3$	$3\sigma v$		
A_1	1	1	1	z	
A_2	1	1	–1	R_z	
E	2	–1	0	$(x,y)(R_x,R_y)$	$(x^2-y^2,xy)\ (x^2,y^2)$

$$\text{No. of } A_1 = \frac{1}{6}[4 + 2 + 6] = 2$$

$$\text{No. of } A_2 = \frac{1}{6}[4 + 2 - 6] = 0$$

$$\text{No. of } E = \frac{1}{6}[8 - 2 - 0] = 1$$

$\therefore$ RR (hybrid orbitals) = $2A_1 + E$

The orbitals corresponding to the symmetry A_1 and E combine to give hybrid orbitals. From the character table, it can be seen that these symmetries represent the s, p_z, p_x and p_y orbitals respectively. Thus, the hybridisation involved in NH_3 is sp^3.

6.2.6 SF_6

The SF_6 molecule belongs to the O_h point group. The various classes of symmetry elements are E, C_3, C_2, C_4, $C_4^2 = C_2$, i, S_4, S_6, σ_h and σ_d. The character representation corresponding to the 6 hybrid orbitals ϕ_1, ϕ_2, ϕ_3, ϕ_4, ϕ_5 and ϕ_6 can be derived as follows.

$$\chi_E = 6 - 0 = 6 \qquad \chi_i = 0 - 0 = 0$$

$$\chi_{C_3} = 0 - 0 = 0 \qquad \chi_{S_4} = 0 - 0 = 0$$

$$\chi_{C_2} = 0 - 0 = 0 \qquad \chi_{S_6} = 0 - 0 = 0$$

$$\chi_{C_4} = 2 - 0 = 2 \qquad \chi_{\sigma_h} = 4 - 0 = 4$$

$\chi_{C_4^2} = 2 - 0 = 2 \qquad \chi_{\sigma_d} = 0 - 0 = 0$

∴ The character representation is

E	C_3	C_2	C_4	C_4^2	i	S_4	S_6	σ_h	σ_d
6	0	0	2	2	0	0	0	4	0

This RR on reduction using the character table gives irreducible representations, $A_{1g} + E_g + T_{1u}$.

A_{1g} corresponds to the most symmetric *s* orbital, E_g corresponds to $d_{x^2-y^2}$ and d_{z^2} orbitals and T_{1u} indicates p_x, p_y and p_z orbitals. Thus, the hybridisation will be sp^3d^2.

Thus, by choosing the hybrid orbitals as the base, RRs can be derived either by the collection of transformation matrices for all symmetry elements of a point group or from the character representation directly by short methods. Such RRs can be expressed as combinations of IRs of the group. From these, the symmetry and nature of hybrid orbitals can be determined.

7

Determination of Symmetry of Vibrational Modes

7.1 General Discussion

All molecules are associated with rotational, translational and vibrational motions due to the displacement of the constituent atoms. These are associated with degrees of freedom that are responsible for their kinetic energy. Among the three degrees of freedom, it is possible to have fundamental bases for translational and rotational movement, therefore, we can derive irreducible representations for them to indicate their symmetries. For vibrational degrees of freedom, a combination of bases has to be considered, that leads to reducible representations which on reduction indicate the symmetry of the vibrational modes of a molecule. Determination of the vibrational symmetry of a molecule basically involves the determination of the symmetry corresponding to the total movement of the molecule and subtraction of the symmetries corresponding to the translational and rotational movements.

The various steps involved in the determination of the symmetry of the vibrational modes of a molecule are:

(i) The point group of the molecule is identified and all symmetry operations for that point group with appropriate classes are listed.

(ii) Three vectors are attached to every atom present in a molecule. These vectors form the basis for the representation corresponding to the total movement of the molecule. The character corresponding to every symmetry operation of the representation for the total movement can be derived by using the following equation.

χ_R = Number of atoms which are not shifted × character corresponding to the same symmetry operation in a 3 × 3 matrix representation

(iii) Using the orthogonality theorem and the character table, the RR can be expressed as a combination of various IRs. From the symmetries corresponding to the total movement of a molecule, the symmetries due to the translational and rotational movements are subtracted in order to arrive at symmetries of the vibrational modes of the molecule.

(iv) From the symmetries of the vibrational modes for simple molecules, the nature of vibration can be identified.

7.2 WORKED EXAMPLES

Example 7.1

Find the symmetry of the vibrational modes of the H_2O molecule.

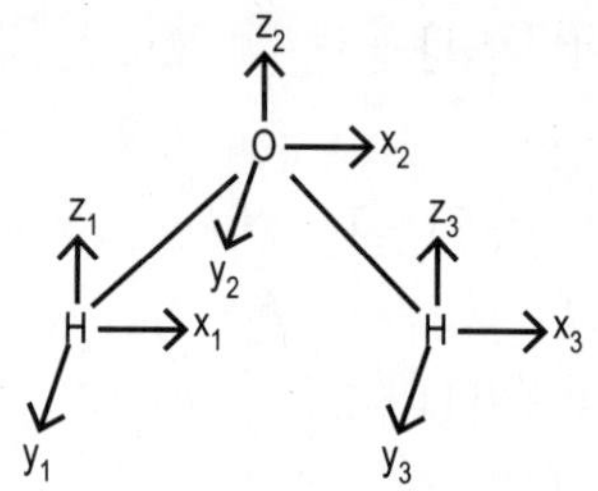

Fig .7.1

The water molecule belongs to the C_{2V} point group and the various symmetry elements present are E, $C_{2(Z)}$, σ_{xz}, σ_{yz}.

Let (x_1,y_1,z_1), (x_2,y_2,z_2) and (x_3,y_3,z_3) be the vectors attached to the $H_{(1)}$, $O_{(2)}$ and $H_{(3)}$ atoms respectively. Total movement of the molecule involves the transformation of all these vectors. Hence all these vectors are taken as bases for the representation corresponding to the total movement of the molecule.

The character corresponding to every symmetry operation of the representation for the total movement can be derived by using the following equation.

χ_R = Number of atoms which are not shifted × character corresponding to the same symmetry operation in a 3 × 3 matrix representation

Thus,

$$\chi_E = 3 \times 3 = 9$$

$$\chi_{C_{2z}} = 1 \times -1 = -1$$

$$\chi_{\sigma_{xz}} = 3 \times 1 = 3$$

$\chi_{\sigma_{yz}} = 1 \times 1 = 1$

The character representation is, 9 –1 3 1.

Using the orthogonality theorem, this can be reduced to the IRs of the group as follows.

C_{2V}	E	C_{2Z}	σ_{xz}	σ_{yz}
$\chi_{\text{Total movement}}$	9	–1	3	1
A_1	1	1	1	1
A_2	1	1	–1	–1
B_1	1	–1	–1	1
B_2	1	–1	1	1

$$\text{No. of } A_1 = \frac{1}{4}[9 - 1 + 3 + 1] = 3$$

$$\text{No. of } A_2 = \frac{1}{4}[9 - 1 - 3 - 1] = 1$$

$$\text{No. of } B_1 = \frac{1}{4}[9 + 1 - 3 + 1] = 2$$

$$\text{No. of } B_2 = \frac{1}{4}[9 + 1 + 3 - 1] = 3$$

∴ The reducible representation corresponding to the total movement of the molecule = $3A_1 + A_2 + 2B_1 + 3B_2$

The symmetries of the translational and rotational modes from the character table = $A_1 + 2B_2 + 2B_1 + A_2$

∴ The symmetry of the vibrational modes = $(3A_1 + A_2 + 2B_1 + 3B_2)$
$- (A_1 + 2B_2 + 2B_1 + A_2)$
$= 2A_1 + B_2$

To find the nature of the vibrational modes of the H_2O molecule,
1. Let us consider the symmetric stretching mode of vibration and subject it to the symmetry operations of the C_{2v} point group (Fig. 7.2).

Fig. 7.2

C_{2V}	E	C_{2Z}	σ_{xz}	σ_{yz}
Symmetric stretching A_1	1	1	1	1

Note: Positive characters indicate that no change has occurred in the direction of oscillation when the respective symmetry operations are performed and vice versa.

The above character representation corresponds to A_1 symmetry. Hence, the symmetric stretching mode of vibration of the H_2O molecule has A_1 symmetry.

2. Let us consider the anti-symmetric stretching mode of vibration (Fig. 7.3).

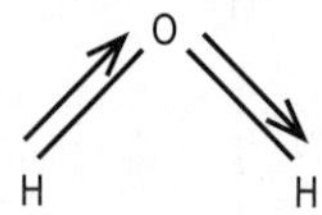

Fig. 7.3

The character representation is given by

E	C_{2z}	σ_{xz}	σ_{yz}
1	–1	1	–1

This character representation corresponds to B_2 symmetry. Thus, the anti-symmetric mode of vibration of the H_2O molecule has B_2 symmetry.

3. Consider the scissoring mode of vibration (Fig.7.4).

Fig. 7.4

The character representation for the above mode is given as,

E	C_{2z}	σ_{xz}	σ_{yz}
1	1	1	1

This corresponds to A_1 symmetry. Therefore, the scissoring mode of vibration of the H_2O molecule has A_1 symmetry.

Example 7.2

Find the symmetry of the vibrational modes of the NH_3 molecule.

The NH_3 molecule belongs to the C_{3V} point group. The various symmetry elements present are E_1, $2C_3$, $3\sigma_v$. Let us assume the total movement of the NH_3 molecule as the base. The character corresponding to the various symmetry operations are derived using the method mentioned above.

$$\chi_E = 4 \times 3 = 12$$

$$\chi_{C_3} = 1 \times 0 = 0$$

$$\chi_{\sigma_v} = 2 \times 1 = 2$$

	E	C_3	σ_v
∴ RR (total movement)	12	0	2

Using the orthogonality theorem and the character table, the RR can be reduced to the irreducible representations of the group.

C_{3V}	E	$2C_3$	$3\sigma_v$
$\chi_{\text{RR Total movement}}$	12	0	2
A_1	1	1	1
A_2	1	1	–1
E	2	–1	0

No. of A_1 $= \frac{1}{6}[12 + 0 + 6] = 3$

No. of A_2 $= \frac{1}{6}[12 + 0 - 6] = 1$

No. of E $= \frac{1}{6}[24 + 0 + 0] = 4$

∴ RR (total movement) $= 3A_1 + A_2 + 4E$

IRs due to rotation and translation modes $= A_1 + A_2 + 2E$

∴ IR due to vibrational modes $= 2A_1 + 2E$

Let us now deduce the nature of the vibrational modes of the NH_3 molecule. There are two modes of vibration having A_1 symmetry. That is, symmetric

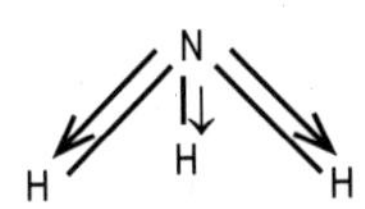

Symmetric stretching

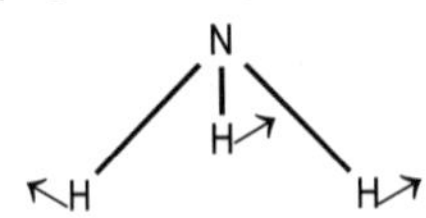

Symmetric bending

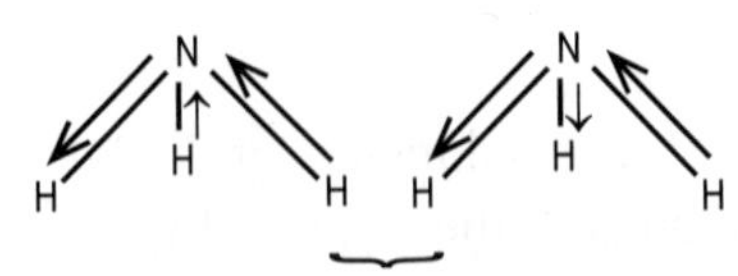

Two fold degenerate anti symmetric stretching

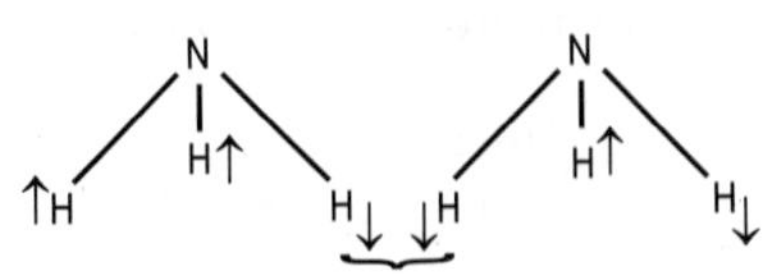

Two fold degenerate anti symmetric bending

Fig. 7.5

stretching and symmetric bending. The character representation for symmetric stretching and bending corresponds to (1, 1, 1) and hence this mode has A_1 symmetry. That is, two A_1 symmetry modes of vibration.

E represents a two fold degenerate vibrational mode. This means that two vibrational modes have the same energy. Therefore, we have accounted for a total of four vibrational modes. Anti-symmetric stretching and anti-symmetric bending modes of vibration belong to E symmetry.

Example 7.3

Find the symmetry of the vibrational modes of the BCl_3 molecule.

The BCl_3 molecule belongs to the D_{3h} point group. The various classes of symmetry elements include E, $2C_3$, $3C_2$, σ_h, $2S_3$ and $3\sigma_v$. Taking total movement of the molecule as the base, the characters for the corresponding reducible representation can be derived as follows.

$\chi_E = 4 \times 3 = 12$ $\qquad \chi_{\sigma_h} = 4 \times 1 = 4$

$\chi_{C_3} = 1 \times 0 = 0$ $\qquad \chi_{\sigma_v} = 2 \times 1 = 2$

$\chi_{C_2} = 2 \times -1 = -2$ $\qquad \chi_{S_3} = 1 \times -2 = -2$

	E	$2C_3$	$3C_2$	σ_h	$2S_3$	$3\sigma_v$
$\therefore \chi$ Total movement	12	0	−2	4	−2	2

The above RR can be reduced to following IRs.

$$\text{No. of } A_1' = 1/12\,[(1\times12\times1)+(2\times0\times1)+(3\times2\times1)+(1\times4\times1)+(2\times-2\times1)+(3\times2\times1)]$$
$$= 1/12\,[12-6+4-4+6]$$
$$= 1$$

$$\text{No. of } A_2' = 1/12\,[(1\times12\times1)+(2\times0\times1)+(3\times-2\times1)+(1\times4\times1)+(2\times-2\times1)+(3\times2\times-1)]$$
$$= 1/12\,[12+0+6+4-4-6]$$
$$= 1$$

$$\text{No. of } E' = 1/12\,[(1\times12\times2)+(2\times0\times-1)+(3\times-2\times0)+(1\times4\times12)+(2\times-2\times-1)+(3\times2\times0)]$$
$$= 1/12\,[24+0+0+8+4+0]$$
$$= 3$$

$$\text{No. of } A_1'' = 1/12\,[(1\times12\times1)+(2\times0\times1)+(3\times-2\times1)+(1\times4\times-1)+(2\times-2\times-1)+(3\times2\times-1)]$$
$$= 1/12\,[12+0-6-4+4-6]$$
$$= 0$$

$$\text{No. of } A_2'' = 1/12\,[(1\times12\times1)+(2\times0\times1)+(3\times-2\times-1)+(1\times4\times-1)+(2\times-2\times-1)+(3\times2\times1)]$$
$$= 1/12\,[12+0+6-4+4+6]$$
$$= 2$$

$$\text{No. of } E'' = 1/12\,[(1\times 12\times 1)+(2\times 0\times 1)+(3\times -2\times 0)+(1\times 4\times -2)+(2\times -2\times 1)+(3\times 2\times 0)]$$
$$= 1/12\,[\,24+0+0-8-4+0\,]$$
$$= 1$$

$\therefore$ RR (total movement) $= A_1' + A_2' + 3E' + 2A_2'' + E''$

IR due to translation and rotation modes $= A_2' + E' + A_2'' + E''$

(from character table)

$\therefore$ The symmetry of the vibrational modes $= A_1' + 2E' + A_2''$

Example 7.4

Find the symmetry of the vibrational modes of the CH_4 molecule.

The CH_4 molecule belongs to the *Td* point group. Various classes of symmetry elements include E, $8C_3$, $3C_2$, $6S_4$ and $6\sigma_d$. The reducible representation for the total movement of the molecule can be derived as follows.

$\chi_E = 5\times 3 = 15$ $\qquad \chi_{S_4} = 1\times -1 = -1$

$\chi_{C_3} = 2\times 0 = 0$ $\qquad \chi_{\sigma_d} = 3\times 1 = 3$

$\chi_{C_2} = 1\times -1 = -1$

	E	$8C_3$	$3C_2$	$6S_4$	$6\sigma_d$
$\therefore$ $RR_{\text{Total movement}}$	15	0	−1	−1	3

The above RR can be reduced to the following IRs.

$$\text{No. of } A_1 = 1/24\,[(1\times 15\times 1)+(8\times 0\times 1)+(3\times -1\times 1)+(6\times -1\times 1)+(6\times 3\times 1)]$$
$$= 1/24\,[\,15+0-3-6+18\,]$$
$$= 1$$

$$\text{No. of } A_2 = 1/24\,[(1\times 15\times 1)+(8\times 0\times 1)+(3\times -1\times 1)+(6\times -1\times -1)+(6\times 3\times -1)]$$
$$= 1/24\,[\,15+0-3+6-18\,]$$
$$= 0$$

$$\text{No. of } E = 1/24\,[(1\times 15\times 2)+(8\times 0\times 1)+(3\times -1\times 2)+(6\times -1\times 0)+(6\times 3\times 0)]$$
$$= 1/24\,[\,30+0+-6+0+0\,]$$
$$= 1$$

$$\text{No. of } T_1 = 1/24\,[(1\times 15\times 3)+(8\times 0\times 0)+(3\times -1\times -1)+(6\times -1\times 1)+(6\times 3\times -1)]$$
$$= 1/24\,[\,30+0-6+0+0\,]$$
$$= 1$$

$$\text{No. of } T_2 = 1/24\,[(1\times 15\times 3)+(8\times 0\times 8)+(3\times -1\times -1)+(6\times -1\times -1)+(6\times 3\times 1)]$$
$$= 1/24\,[\,45+0+3+6+18\,]$$
$$= 3$$

$\therefore$ RR (total movement) = $A_1 + E + T_1 + 3T_2$

IR due to translation and rotation modes = $T_1 + T_2$

(from character table)

$\therefore$ Symmetry of the vibrational modes = $A_1 + E + 2T_2$

Example 7.5

Find the symmetry of the vibrational modes of the XeF_4 molecule.

The XeF_4 molecule belongs to the D_{4h} point group. The various symmetry operations of the group are E, $2C_4$, $C_4^2 = C_2$, $2C_2$, $2C_2'$, i, $2S_4$, σ_h, $2\sigma_v$ and $2\sigma_d$. The reducible representation corresponding to the total movement of the molecule can be derived as follows.

$\chi_E = 5 \times 3 = 15$ $\quad\quad \chi_i = 1 \times -3 = -3$

$\chi_{C_4} = 1 \times 1 = 1$ $\quad\quad \chi_{\sigma_h} = 5 \times 1 = 5$

$\chi_{C_4^2 - C_2} = 1 \times -1 = -1$ $\quad\quad \chi_{\sigma_v} = 1 \times 1 - 1$

$\chi_{C_2} = 1 \times -1 = -1$ $\quad\quad \chi_{S_4} = 1 \times -1 = 1$

$\chi_{C_2'} = 3 \times -1 = -3$ $\quad\quad \chi_{\sigma_d} = 3 \times 1 = 3$

	E	$2C_4$	$C_4^2 = C_2$	$2C_2$	$2C_2'$	i	$2S_4$	σ_h	$2\sigma_v$	$2\sigma_d$
$\therefore$ RR $_{\text{Total movement}}$	15	1	–1	–1	–3	–3	–1	5	1	3

On reduction using the character table,

$$\text{No. of } A_{1g} = 1/16\,[(1 \times 15 \times 1) + (2 \times 1 \times 1) + (1 \times -1 \times 1) + (2 \times -1 \times 1) + (2 \times -3 \times 1) + (1 \times -3 \times 1) + (2 \times -1 \times 1) + (1 \times 5 \times 1) + (2 \times 1 \times 1) + (2 \times 3 \times 1)]$$
$$= 1/16\,[15 + 2 - 1 - 2 - 6 - 3 - 2 + 5 + 2 + 6]$$
$$= 1$$

$$\text{No. of } A_{1u} = 1/16\,[(1 \times 15 \times 1) + (2 \times 1 \times 1) + (1 \times -1 \times 1) + (2 \times -1 \times 1) + (2 \times -3 \times 1) + (1 \times -3 \times -1) + (2 \times -1 \times -1) + (1 \times 5 \times -1) + (2 \times 1 \times -1) + (2 \times 3 \times -1)]$$
$$= 1/16\,[15 + 2 - 1 - 2 - 6 + 3 + 2 - 5 - 2 - 6]$$
$$= 0$$

$$\text{No. of } A_{2g} = 1/16\,[(1 \times 15 \times 1) + (2 \times 1 \times 1) + (1 \times -1 \times 1) + (2 \times -1 \times -1) + (2 \times -3 \times -1) + (1 \times -3 \times 1) + (2 \times -1 \times 1) + (1 \times 5 \times 1) + (2 \times 3 \times -1) + (2 \times 3 \times 1)]$$
$$= 1/16\,[15 + 2 - 1 + 2 + 6 - 3 - 2 + 5 - 2 - 6]$$
$$= 1$$

Similarly, No. of A_{2u} = 2

No. of B_{1g} = 1

No. of B_{2g} = 1

No. of B_{2u} = 1

No. of E_g = 1
No. of E_u = 3

$\therefore$ RR (total movement) = $A_{1g} + B_{1g} + 2A_{2u} + A_{2g} + B_{2g} + B_{2u} + E_g + 3E_u$
The IRs due to translation and rotation modes = $A_{2u} + A_{2g} + E_g + E_u$
(from character table)
$\therefore$ symmetry of the vibrational modes = $A_{1g} + B_{1g} + A_{2u} + B_{2g} + B_{2u} + 2E_u$

This chapter thus summarises the Cartesian method for determining the symmetry of the total movement of the molecule. From that, the symmetry of the vibrational modes for the same molecule can be deduced. The method has been explained with specific examples. For simple molecules, even the nature of hybridisation can be derived from the symmetry of hybridisation.

8

Infrared and Raman Activity of Molecular Vibrations

8.1 Introduction

When molecules are exposed to infrared radiations (IR) having frequencies that are equal to the fundamental vibrational frequencies of the molecule, vibrational transitions can take place by a dipole mechanism. The probability of such a transition occurring, however, depends on the relationship between the molecule's electric dipole moment and the wave functions of the ground state and the higher vibrational states. According to quantum mechanics, the probability of the transition is proportional to the following integral,

$$\int \psi_n^* \, \mu \, \psi_m \, dr$$

where, ψ_n and ψ_m represent the wave functions of the ground state and the excited state respectively and μ, the dipole moment operator.

The transition is forbidden if the integral $\int \psi_n^* \, \mu \, \psi_m \, dr$ vanishes and it is allowed if the integral does not vanish. Quantum mechanical evaluation of the integral is fairly difficult.

Using group theory, a simpler method can be adopted. The symmetries of the ground state, the excited state and the dipole moment operator are identified and the overall direct product of all the three is determined. If the direct product transforms to the most symmetric representation of a group, it is equivalent to a non-vanishing integral and vice versa. Using this method, the allowedness of spectral transitions can be predicted.

8.2 DIRECT PRODUCT REPRESENTATION

If irreducible representations corresponding to simple functions or bases are known, representations for the products of two or more functions (bases) can be derived from the direct product of the respective irreducible representations. We come across such situations when we deal with the overall wave function expressed as the product of wave functions relating to the electronic, rotational and vibrational motions of the molecule.

Let G_A and G_B be the irreducible representations of two functions f_A and f_B. Further, let A and B be the matrices for the two corresponding IRs of certain symmetry operations of the group. For the same symmetry operation, the matrix corresponding to the product of functions f_A and f_B is given by the direct product of A with B represented as A × B. The direct product is different from matrix multiplication and it can be derived as follows.

Let

$$A = \begin{pmatrix} a_1 & a_2 \\ a_3 & a_4 \end{pmatrix} \quad \text{and } B = \begin{pmatrix} b_1 & b_2 \\ b_3 & b_4 \end{pmatrix}$$

$$\chi_A = (a_1 + a_4) \qquad \chi_B = (b_1 + b_4)$$

Direct product of A with B is given as

$$A \times B = \begin{pmatrix} a_1 & a_2 \\ a_3 & a_4 \end{pmatrix} \times \begin{pmatrix} b_1 & b_2 \\ b_3 & b_4 \end{pmatrix}$$

$$= \begin{pmatrix} a_1b_1 & a_1b_2 & a_2b_1 & a_2b_2 \\ a_1b_3 & a_1b_4 & a_2b_3 & a_2b_4 \\ a_3b_1 & a_3b_2 & a_4b_1 & a_4b_2 \\ a_3b_3 & a_3b_4 & a_4b_3 & a_4b_4 \end{pmatrix}$$

The character corresponding to the direct product is given by,

$$\chi_{AB} = a_1b_1 + a_1b_4 + a_4b_1 + a_4b_4$$
$$= a_1(b_1 + b_4) + a_4(b_1 + b_4)$$
$$= (a_1 + a_4)(b_1 + b_4)$$

The direct product matrix has a higher order than the matrices A or B.

Since $(a_1 + a_4) = \chi_A$, and $(b_1 + b_4) = \chi_B$,

$$\chi_{A \times B} = \chi_A \times \chi_B$$

That is, the character of the direct product of two matrices is equal to the product of the characters of A and B.

A collection of direct product matrices for all the symmetry operations of the group form a representation corresponding to the product function.

8.3 DETERMINATION OF DIRECT PRODUCT

Let us consider $\vec{Z}$ and $\vec{Y}$ belonging to the C_{2V} point group. $\vec{Z}$ has A_1 symmetry and $\vec{Y}$ has B_2 symmetry. The symmetry of the product vector $(\overrightarrow{ZY})$ can be obtained from their direct product as shown below.

Base	**Symmetry**	E	$C_{2(Z)}$	σ_{xz}	σ_{yz}
$\vec{Z}$	A_1	1	1	1	1
$\vec{Y}$	B_2	1	–1	–1	1
$\vec{Z}\times\vec{Y}$	$A_1 \times B_2$	1 × 1	1 × –1	1 × –1	1 × 1
$\vec{Z}\times\vec{Y}$	B_2	1	–1	–1	1

$$\therefore A_1 \times B_2 = B_2$$

Thus, the symmetry of $\overrightarrow{ZY}$ is B_2. Similarly, direct product of other one-dimensional representations can be determined.

Let us apply the same procedure for determining the direct product of two degenerate representations. For example, the direct product of E with E in the C_{3V} point group can be derived as follows.

C_{3V}	E	$2C_3$	$3\sigma_V$
E	2	–1	0
$E \times E$	2 × 2	–1 × –1	0 × 0
RR	4	1	0

The direct product above gives a reducible representation. This reducible representation can be reduced to various irreducible representations of a group using the character table.

$$\therefore E \times E = A_1 + A_2 + E$$

Thus, the direct product of two one-dimensional representations remains a single irreducible representation whereas the direct product of multi-dimensional representations yields a sum of the various IRs of the group.

8.4 INFRARED ACTIVITY OF VIBRATIONAL MODES

It is understood that only when a vibration is capable of changing the dipole moment of the molecule during oscillation, it can be IR active. To identify such IR-active vibrations, the overall direct product of the symmetries of the ground state, the dipole moment operator and the excited state has to be determined. If the overall direct product is the most symmetric representation of a group, the vibration is IR active and vice versa. If the direct product transforms to the most symmetric representation of a group, it indicates that the symmetry of the molecule is not disturbed by the vibrational transition of the molecule taking place in the presence of IR radiations. That is, radiative vibrational absorption does not influence the symmetrical environment of the molecule.

As the symmetry of the ground state corresponds to the most symmetric representation of the group, only when the symmetry of the vibrational mode is the same as that of any one dipole moment component operator, the overall direct product will become the most symmetric representation of the group and become IR-active. Thus, IR-active vibrational modes will possess the symmetry of any one component of a dipole moment operator.

For example, the water molecule has three vibrational modes, out of which two modes have A_1 symmetry and one mode has B_2 symmetry. It can be seen from the character table that the dipole moment component μ_z has A_1 symmetry and μ_y has B_2 symmetry. The symmetry of the ground state is generally the most symmetric representation of the group.

Hence, the overall direct product corresponding to the A_1 vibration is,

$$A_1 \times A_1 \times A_1 = A_1 \qquad (\because \mu_z = A_1)$$

Similarly, the overall direct product for the B_2 mode of vibration is,

$$A_1 \times B_2 \times B_2 = A_1 \qquad (\because \mu_y = B_2)$$

Since the most symmetric representation in both cases is obtained as the overall direct product, both are expected to be IR active.

The NH_3 molecule has [(3 × 4) – 6 = 6] six vibrational degrees of freedom having A_1 and E symmetries. The dipole moment component operator μ_z has A_1 symmetry and the dipole moment components μ_x and μ_y together have E symmetry.

For vibrational modes of A_1 symmetry, the overall direct product is

$$A_1 \times A_1 \times A_1 = A_1 \qquad (\because \mu_z = A_1)$$

Similarly for the E mode of vibration, the overall direct product is

$$A_1 \times E \times E = A_1 + A_2 + E \qquad (\mu_x, \mu_y = E)$$

Since A_1 is present as a direct product in both cases, the E and the A_1 modes of vibrations are IR active.

The symmetry of the vibrational modes of XeF_4 are A_{1g}, B_{1g}, B_{2g}, B_{2u} and E_u. The symmetry of the ground state of XeF_4 is A_{1g}. The μ_z operator has A_{2u} symmetry and the μ_x and μ_y operators together have E_{1u} symmetry. The overall direct product will be transformed to the most symmetric representation, that is, the A_{1g} of the group, only for the vibrational modes of symmetry A_{2u} and E_{1u}. They are thus, IR-active. The other modes of vibrations are IR-inactive.

8.5 Raman Activity of Vibrational Modes

IR-active vibrations involve a change in dipole moment during oscillation. The condition for Raman activity is that the vibrations should involve a change in polarisability or induced dipole moment during oscillation. To determine such Raman active vibrations, the symmetry of the ground state, the symmetry of the excited state and the symmetry of the polarisability operator should be known and their overall direct product has to be determined. The symmetry of the ground state is as usual the most symmetric representation of the group and the symmetry of the excited state is the symmetry of the vibrational mode. The polarisability or the induced dipole moment operator has many components which are given as $\hat{\mu}_{x^2}$, $\hat{\mu}_{y^2}$, $\hat{\mu}_{z^2}$, $\hat{\mu}_{xy}$, $\hat{\mu}_{xz}$, $\hat{\mu}_{yz}$ and so forth. The symmetries of these induced dipole moment components are the symmetries of the corresponding product vectors, that is, the symmetries of x^2, xy, xz, y^2, yz and z^2. If the symmetry of any one component of the induced dipole moment operator remains the same as that of the vibrational mode, the overall direct product of the symmetry of the ground state, the excited state and the operator will be the most symmetric representation of the group. Therefore, such vibrations will be Raman-active and vice versa.

For example, the H_2O molecule has A_1 and B_2 symmetric vibrational modes. The character table indicates the symmetry of the polarisability operator components $\hat{\mu}_{x^2}$, $\hat{\mu}_{y^2}$ and $\hat{\mu}_{z^2}$ as A_1 and the symmetry of the component $\hat{\mu}_{xy}$ as B_2. Thus, for the A_1 mode of vibration, the overall direct product $= A_1 \times A_1 \times A_1 = A_1$ and for the B_2 mode of vibration, the overall direct product $= A_1 \times B_2 \times B_2 = A_1$. Therefore, both these vibrational modes are Raman-active.

The NH_3 molecule has vibrations with symmetries A_1 and E. From the character table it can be seen that the polarisability operator components $\hat{\mu}_{x^2}$, $\hat{\mu}_{y^2}$ and $\hat{\mu}_{z^2}$ have A_1 symmetry and the component $\hat{\mu}_{xy}$ has E symmetry. Hence, both the direct products will lead to the most symmetric representation of the group and hence will be Raman-active.

8.6 Mutual Exclusion Principle

It is observed that in the case of molecules having a centre of symmetry, the infrared- active modes are Raman-inactive and Raman-active modes are IR-inactive.

This observation gives rise to the mutual exclusion principle. It states that, *in the case of centrosymmetric molecules, the IR-active vibrational modes are Raman-inactive and the Raman-active vibrational modes are IR-inactive*. This principle can also be deduced using the direct product concept.

The dipole moment operator components have the symmetries of certain coordinates x, y and z. These coordinates are anti-symmetric to the inversion operation (*i*) and hence have the 'u' symmetry. When the vibrational modes are also anti-symmetric to inversion or possess 'u' symmetry, then the overall direct product will become symmetric (u × u = g) and be IR-active. If the vibrational modes are symmetric to inversion (g), then the direct product will become (g × u = u) anti-symmetric and therefore, such vibrations will be IR-inactive. Thus, in the case of molecules with a centre of symmetry, the vibrational modes symmetric to '*i*' (g modes) are IR-inactive whereas the vibrational modes that are anti-symmetric to '*i*' (u mode) are IR-active.

The polarisability operator components have the symmetry of the binary products of the cartesian coordinates such as x^2, y^2, z^2, xy, yz and xz. These components are symmetric to the inversion operation and therefore have 'g' symmetry. The vibrational modes that are symmetric to inversion (g modes) will lead to symmetric direct products (g × g = g) accounting for the Raman activity of such vibrational modes. If the vibrational modes are anti-symmetric to inversion, that is, those that possess 'u' symmetry, the direct product with any component of the polarisability operator will become anti-symmetric ('u' symmetric) and will thus be Raman-inactive. Thus, vibrations of 'g' modes are Raman-active and vibrations of 'u' modes are Raman-inactive which is the opposite of the predictions made for IR activity. Since the terms 'g' and 'u' have relevance only with molecules that can undergo the inversion operation, the mutual exclusion principle is applicable only for centro-symmetric molecules.

8.7 IR AND RAMAN ACTIVITY OF OVERTONES AND COMBINATION BANDS

In the earlier sections, the IR and Raman activity of fundamental vibrational absorptions has been discussed. Both in the IR and the Raman spectra, apart from the fundamental absorption, overtones and combination bands are also present. The overtones are due to vibrational transitions from v = 0 to v = 2, 3 and other higher levels. The combination bands have their frequency almost equal to the sum of two different fundamental absorption frequencies. The conditions for the appearance of overtones and combination bands in the IR and Raman spectra are the same as that of fundamental vibrational absorptions. That is, to be IR-active, the symmetry of the overtone or combination band should correspond to the symmetry of any one component of the dipole moment operator and to be Raman-active, the symmetry of overtone or combination band should correspond to the symmetry of any one component of the polarisability operator.

However, the symmetries of the overtones and combination bands are derived from their respective direct products.

For example, if a vibrational mode has A_2 symmetry, its first overtone symmetry is A_1 which is obtained by the following direct product $A_2 \times A_2 = A_1$.

Similarly, a combination band is due to $\nu_1 + \nu_2$ (ν_1 and ν_2 represent the oscillator frequency) where, ν_1 has A_1 symmetry and ν_2 has B_1 symmetry, and the symmetry of the combination band is $A_1 \times B_1 = B_1$.

IR and Raman spectra are generally used to predict the structures of molecules. In the last chapter, the determination of symmetries of vibrational modes of poly atomic molecules using group theory was discussed. In this chapter, the prediction of the IR and Raman activity of various vibrational modes of a molecule using group theory, have been dealt with. Thus, overall, group theory helps in predicting the number of fundamental bands, combination and difference bands in both IR and Raman spectra. The comparison between the observed IR and Raman spectra and the theoretically predicted activity leads to the structural assignment. For example, consider a molecule AB_3, it can have any one of the following geometries.

(i) Planar – D_{3h}
(ii) Pyramidal – C_{3v}
(iii) T-shaped – C_{2v}

For every geometry, the permitted fundamentals in IR and Raman spectra can be determined using the above mentioned methods of group theory. The experimental IR and Raman bands are fitted with these theoretical predictions and the structure corresponding to the best fit is taken as the correct structure. Tables 8.1 and 8.2 illustrate this method.

Table 8.1

Structure	Point group	IR (theoretical prediction)	Raman (theoretical prediction)
Planar	D_{3h}	3	3
Pyramidal	C_{3v}	4	4
T-Shaped	C_{2v}	6	6

Table 8.2

Molecule	Observed IR bands	Observed Raman bands	Predicted geometry
PCl_3	4	4	Pyramidal
BF_3	3	3	Planar
ClF_3	6	6	T-Shaped

9

Selection Rules for Electronic Transitions

9.1 Introduction

Molecules can have many types of energy. They have rotational energy by virtue of bodily rotation, vibrational energy due to oscillatory movement, electronic energy, nuclear energy and spin energy. All these energy levels are quantised and the transitions taking place among respective energy levels in the presence of electromagnetic radiation lead to various spectra. In the last chapter, the transition probabilities for IR and Raman spectra were dealt with applying group theoretical methods and the present chapter deals with electronic transitions.

The excitation of a valence electron involves the movement of electronic charges in a molecule. The consequent changes occurring in the electric dipoles, give rise to a spectrum due to their interaction with the undulatory electric field of radiation.

9.2 Electronic Transitions

When a molecule is exposed to visible or near-UV radiation, it can undergo electronic transitions by dipole mechanism, provided the transition is symmetrically permitted or allowed. The intensity of the transition is proportional to the integral $\int \chi_g^* \chi_\mu \chi_e \, dr$ where, χ_g and χ_e represent the electronic wave function corresponding to the ground electronic state and the excited electronic state respectively. χ_μ corresponds to the dipole moment vector. If the integral vanishes, the transition is symmetrically forbidden and vice-versa.

We have discussed in the last chapter that the evaluation of the integral is cumbersome and assessment of the conditions for the transition can be achieved by a simple method using group theory. This method involves the determination of the overall direct product of the symmetry of the ground electronic state, the symmetry of the dipole moment operator and the symmetry of the excited electronic state. If the overall direct product transforms to the most symmetric representation of the group, it indicates that the symmetry of the molecule is not disturbed by the electronic transition taking place and hence it is allowed and vice versa. Thus, to identify the symmetrically allowed electronic transitions, the symmetry of various electronic states has to be determined.

The types of molecular orbitals present in a molecule are generally σ, π, *n*, π* and σ*. Saturated compounds have only σ and σ* orbitals whereas unsaturated compounds have σ, π, π* and σ* orbitals. When a heteroatom is present along with unsaturation in a compound, it will have σ, π, *n*, π* and σ* orbitals. Thus, the various electronic transitions that are possible are, $\sigma \rightarrow \sigma^*$, $\pi \rightarrow \pi^*$, $n \rightarrow \pi^*$ and $n \rightarrow \sigma^*$. Out of these transitions, the $\sigma \rightarrow \sigma^*$ involves a large amount of energy and hence occurs in the far-UV region whereas the remaining transitions occur either in the visible or near-UV region depending on the nature of molecule.

9.3 SYMMETRY OF MOLECULAR ORBITALS AND ELECTRONIC STATES OF FORMALDEHYDE

Formaldehyde has three σ bonding molecular orbitals ($2\sigma_{C-H}$ and $1\sigma_{C-0}$) and the corresponding three σ* anti-bonding molecular orbitals. Apart from this, it has a C–O π orbital, the π* anti-bonding orbital and also two sets of orbitals on the oxygen atom that accommodate the non-bonding electrons. The molecule belongs to the C_{2V} point group and the symmetry operations are E, $C_{2(z)}$, σ_{xz} and σ_{yz}.

The symmetries of the σ_{C-0}, π_{C-0}, π^*_{C-0} and the non-bonding orbitals can be derived by studying how the orbitals get transformed when the four symmetry operations of this group are performed on them.

Base orbital	E	$C_{2(Z)}$	σ_{xz}	σ_{yz}	Symmetry
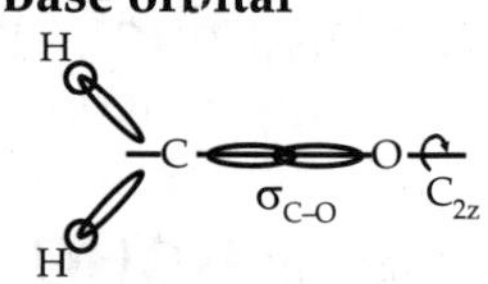	1	1	1	1	A_1
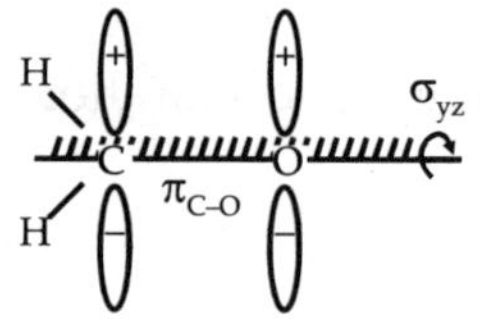	1	–1	–1	1	B_1

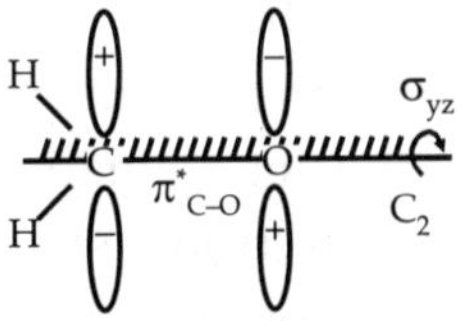

1   −1   −1   1 B_1

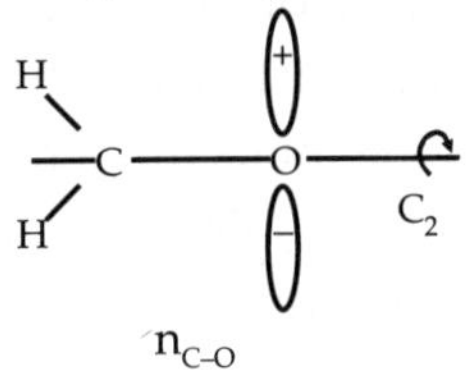

1   −1   1   −1 B_2

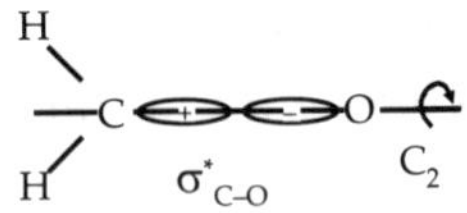

1   1   1   1 A_1

Note: The positive characters indicate that no transformation has occurred to the orbital when the respective symmetry operation has been performed and the negative characters suggest the reversal of the symmetry signs of the orbital after the symmetry operations have been performed.

The ground state electronic configuration of the molecule is $(\sigma_{CO})^2(\pi_{CO})^2 n^2$. The symmetry of this ground state is obtained from the overall direct product of the symmetries of all the orbitals involved. That is,

$(A_1)^2 \times (B_1)^2 \times (B_2)^2 = A_1 \times A_1 \times A_1 = A_1$

If an electronic transition takes place from the π to the π^* orbital, the electronic configuration will be $(\sigma_{CO})^2(\pi_{CO})^1 n^2(\pi^*_{CO})^1$. The symmetry of the $\pi \rightarrow \pi^*$ excited state is given by the following direct product.

$(A_1)^2 \times B_1 \times (B_1)^2 \times B_1 = A_1 \times A_1 \times A_1 = A_1$

Similarly, for a $n - \pi^*$ transition, the electronic configuration is $(\sigma_{CO})^2(\pi_{CO})^2 n^1(\pi^*_{CO})^1$ and the respective symmetry is given by,

$(A_1)^2 \times (B_1)^2 \times B_2 \times B_2 = A_1 \times A_1 \times B_2 \times B_1 = A_2$

For a $n - \sigma^*$ transition, the electronic configuration is $(\sigma_{CO})^2(\pi_{CO})^2 n^1(\sigma^*_{CO})^1$ and the symmetry is $(A_1)^2 \times (B_1)^2 \times B_2 \times A_1 = B_2$.

9.4 SELECTION RULES FOR ELECTRONIC TRANSITIONS OF HCHO

The symmetries of various electronic states associated with ground vibrational states are given by the following direct products.

Ground electron state $= A_1 \times A_1 = A_1$

$\pi - \pi^*$ state $= A_1 \times A_1 = A_1$

n – π^* state $= A_2 \times A_1 = A_2$
n – σ^* state $= B_2 \times A_1 = B_2$

The symmetries of the three dipole moment components are A_1, B_1 and B_2. As discussed under IR and Raman spectra, the condition for an electronic transition to appear in the spectrum is that there should be at least one dipole moment component with the same symmetry as the excited electronic state. By applying this condition, it can be understood that the $\pi - \pi^*$ transition and the n – σ^* transition are symmetrically allowed for formaldehyde. Since none of the dipole moment components have the A_2 symmetry, the n – π^* transition is forbidden. However such n – π^* transitions, if coupled with a vibrational transition of different symmetry so that the overall direct product transforms to the next symmetric representation of the group, become symmetrically allowed. Thus, the n – π^* transition appears with a low intensity in the spectrum whereas the $\pi - \pi^*$ and the n – σ^* bands appear with higher intensity in the spectrum.

Thus, using group theory, the symmetry of various electronic states and the probability of spectral transitions from the ground state to various electronic states can be derived. Such studies help in the interpretation of the electronic spectra of molecules.

Appendix I

Diagonalisation of Matrices

A square matrix **A** of order *n*, with *n* linearly independent eigen vectors can be diagonalised by a similarity transformation done with **B** $(\mathbf{B}^{-1}\,\mathbf{A}\,\mathbf{B})$ where **B** is the matrix whose columns are the eigen vectors of **A**. Hence, to diagonalise a matrix, eigen vectors and eigen values should be evaluated for the matrix.

Eigen values and Eigen vectors

Suppose a vector $\mathbf{X} = \begin{pmatrix} x_1 \\ x_2 \\ x_3 \end{pmatrix}$ can be transformed to another

vector **Y** by means of a matrix **A**, we get

$$\mathbf{Y} = \mathbf{AX} \qquad \text{----------} \ (1)$$

where $\mathbf{A} = \begin{pmatrix} a_{11} & a_{12} & a_{13} \\ a_{21} & a_{22} & a_{23} \\ a_{31} & a_{32} & a_{33} \end{pmatrix}$

Parellel to the vector **X** , if we have

$$\mathbf{Y} = \lambda\mathbf{X} \qquad \text{----------} \ (2)$$

where λ is a scalar, then, combining equations (1) and (2) we have

$$\mathbf{A}\,\mathbf{X} = \lambda\mathbf{X} \qquad \text{----------} \ (3)$$

In the above equation, vector **X** is termed as an eigen vector of matrix **A** or an eigen vector of the given transformation and hence the scalar λ is termed as the eigen value associated with the eigen vector **X**.

Computation of eigen values and eigen vectors

Equation (3) can be written as

$$(\mathbf{A} - \lambda\,\mathbf{I})\ \mathbf{X} = \mathbf{0} \qquad \text{----------} \ (4)$$

where **I** is the unit matrix of the same order as that of **A**.

If $\mathbf{A} = \begin{pmatrix} a_{11} & a_{12} & \cdots\cdots & a_{1n} \\ a_{21} & a_{22} & \cdots\cdots & a_{2n} \\ \cdots & \cdots & \cdots & \cdots \\ a_{n1} & a_{n2} & \cdots\cdots & a_{nn} \end{pmatrix}$ and $\mathbf{I} = \begin{pmatrix} 1\ 0\ 0\ 0 & \cdots\cdots & 0 \\ 0\ 1\ 0\ 0 & \cdots\cdots & 0 \\ \cdots & \cdots & \cdots \\ 0\ 0\ 0\ 0 & \cdots\cdots & 1 \end{pmatrix}$

then $\mathbf{X} = \begin{pmatrix} x_1 \\ x_2 \\ \cdots \\ x_n \end{pmatrix}$

Substituting $\mathbf{A}$, $\mathbf{I}$ and $\mathbf{X}$ in equation (4), we get

$$\begin{pmatrix} a_{11}-\lambda & a_{12} & \cdots\cdots & a_{1n} \\ a_{21} & a_{22}-\lambda & \cdots\cdots & a_{2n} \\ \cdots & \cdots & \cdots & \cdots \\ a_{n1} & a_{n2} & \cdots\cdots & a_{3n}-\lambda \end{pmatrix} \begin{pmatrix} x_1 \\ x_2 \\ \cdots \\ x_n \end{pmatrix} = 0 \quad \text{---------- (5)}$$

$$\begin{aligned} \text{That is, } (a_{11}-\lambda)\,x_1 &+ a_{12}\,x_2 + \cdots\cdots + a_{1n}\,x_n = 0 \\ a_{21} - x_1 &+ (a_{21}-\lambda)\,x_2 + \cdots\cdots + a_{2n}\,x_n = 0 \\ &\cdots\cdots\cdots\cdots\cdots\cdots \quad \text{---------- (6)} \\ (a_{n1}-\lambda)\,x_1 &+ a_{n2}\,x_2 + \cdots\cdots + (a_{n1}-\lambda)\,x_n = 0 \end{aligned}$$

Equation (6) is a system of *n* simultaneous linear algebraic equations for the *n* unknown $x_1, x_2, \ldots x_n$. The necessary condition for the existence of a non-trivial solution to such a system of homogenous equations is the vanishing of the determinant of the coefficients.

$$\text{That is,} \quad \det(\mathbf{A} - \lambda\mathbf{X}) = \begin{vmatrix} a_{11}-\lambda & a_{12} & \cdots\cdots & a_{1n} \\ a_{21} & a_{21}-\lambda & \cdots\cdots & a_{2n} \\ \cdots & \cdots & \cdots & \cdots \\ a_{n1} & a_{n2} & & a_{n1}-\lambda \end{vmatrix} = 0 \quad \text{---------- (7)}$$

Equation (7) is referred to as a secular equation associated with the matrix **A**. This equation is a polynomial equation of degree *n* and therefore possesses precisely *n* roots $\lambda_1, \lambda_2 \cdots\cdots \lambda_n$. Corresponding to each of these λ values, the system of equation (6) has a non trivial solution for eigen vectors.

Example: Let us digonalise the matrix $\mathbf{A} = \begin{pmatrix} 1 & 1 & 3 \\ 1 & 5 & 1 \\ 3 & 1 & 1 \end{pmatrix}$

The secular equation is $\begin{vmatrix} 1-\lambda & 1 & 3 \\ 1 & 5-\lambda & 1 \\ 3 & 1 & 1-\lambda \end{vmatrix} = 0$

That is, $(1-\lambda)[(5-\lambda)(1-\lambda)-1]-1[(1-\lambda)-3]+3[1-3(5-\lambda)]=0$

On simplifying, we get $-\lambda^3+7\lambda^2-36=0$ or $\lambda^3-7\lambda^2+36=0$

Therefore, $\lambda = -2, 3, 6$.

Eigen vectors can be derived by substituting different λ values in the following equation.

$$\begin{pmatrix} 1-\lambda & 1 & 3 \\ 1 & 5-\lambda & 1 \\ 3 & 1 & 1-\lambda \end{pmatrix} \begin{pmatrix} x_1 \\ x_2 \\ x_3 \end{pmatrix} = 0$$

Thus, eigen vectors corresponding to the eigen values –2, 3, 6 are $x_1 = (-1, 0, 1)$, $x_2 = (1, -1, 1)$ and $x_3 = (1, 2, 1)$. Combining x_1, x_2 and x_3 as column elements, matrix **B** can be derived. If a similarity transformation is done with respect to this on matrix **A**, the diagonalised matrix **D** can be obtained.

Therefore $\mathbf{B} = \begin{pmatrix} -1 & 1 & 1 \\ 0 & -1 & 2 \\ 1 & 1 & 1 \end{pmatrix}$

Adjoint of $\mathbf{B} = \begin{pmatrix} -3 & 0 & 3 \\ 2 & -2 & 2 \\ 1 & 2 & 1 \end{pmatrix}$ and $\det \mathbf{B} = 6$

$$\mathbf{B}^{-1} = \text{Adj}\,\mathbf{B} / \det \mathbf{B}$$

$$= \begin{pmatrix} -1/2 & 0 & 1/2 \\ 1/3 & -1/3 & 1/3 \\ 1/6 & 1/3 & 1/6 \end{pmatrix}$$

thus $\mathbf{D} = \mathbf{B}^{-1}\mathbf{A}\mathbf{B} = \begin{pmatrix} -1/2 & 0 & 1/2 \\ 1/3 & -1/3 & 1/3 \\ 1/6 & 1/3 & 1/6 \end{pmatrix} \begin{pmatrix} 1 & 1 & 3 \\ 1 & 5 & 1 \\ 3 & 1 & 1 \end{pmatrix} \begin{pmatrix} -1 & 1 & 1 \\ 0 & -1 & 2 \\ 1 & 1 & 1 \end{pmatrix}$

$\mathbf{D} = \begin{pmatrix} -2 & 0 & 0 \\ 0 & 3 & 0 \\ 0 & 0 & 6 \end{pmatrix}$ is a diagonalised matrix of **A** having eigen values as the diagonal elements.

Appendix II

Character Tables of Symmetry Groups

1. The Nonaxial Groups

C_1	E
A	1

C_s	E	σ_h		
A'	1	1	x, y, R_z	x^2, y^2, z^2, xy
A''	1	–1	z, R_x, R_y	yz, xz

C_i	E	i		
A_g	1	1	R_x, R_y, R_z	$x^2, y^2, z^2, xy, xz, yz$
A_u	1	–1	x, y, z	

2. The C_n Groups

C_2	E	C_2		
A	1	1	z, R_z	x^2, y^2, z^2, xy
B	1	–1	x, y, R_x, R_y	yz, xz

C_3	E	C_3	C_3^2		$\varepsilon = \exp(2\pi i/3)$
A	1	1	1	z, R_z	$x^2 + y^2, z^2$
E	$\begin{Bmatrix} 1 \\ 1 \end{Bmatrix}$	$\begin{matrix} \varepsilon \\ \varepsilon^* \end{matrix}$	$\begin{matrix} \varepsilon^* \\ \varepsilon \end{matrix}$	$(x,y)(R_x, R_y)$	$(x^2 - y^2, xy)(yz,xz)$

C_4	E	C_4	C_2	C_4^3		
A	1	1	1	1	z, R_z	$x^2 + y^2, z^2$
B	1	–1	1	–1		$x^2 - y^2, xy$
E	1	i	–1	$-i$	$(x,y)(R_x, R_y)$	(yz, xz)
	1	$-i$	–1	i		

C_5	E	C_5	C_5^2	C_5^3	C_5^4		$\varepsilon = \exp(2\pi i/5)$
A	1	1	1	1	1	z, R_z	$x^2 + y^2, z^2$
E_1	1	ε	ε^2	ε^{2*}	ε^*	$(x, y)(R_x, R_y)$	(yz, xz)
	1	ε^*	ε^{2*}	ε^2	ε		
E_2	1	ε^2	ε^*	ε	ε^{2*}		$(x^2 - y^2, xy)$
	1	ε^{2*}	ε	ε^*	ε^2		

C_6	E	C_6	C_3	C_2	C_3^2	C_6^5		$\varepsilon = \exp(2\pi i/6)$
A	1	1	1	1	1	1	z, R_z	$x^2 + y^2, z^2$
B	1	–1	1	–1	1	–1		
E_1	1	ε	$-\varepsilon^*$	–1	$-\varepsilon$	ε^*	(x, y)	(xz, yz)
	1	ε^*	$-\varepsilon$	–1	$-\varepsilon^*$	ε	(R_x, R_y)	
E_2	1	$-\varepsilon^*$	$-\varepsilon$	1	$-\varepsilon^*$	$-\varepsilon$		$(x^2 - y^2, xy)$
	1	$-\varepsilon$	$-\varepsilon^*$	1	$-\varepsilon$	$-\varepsilon^*$		

C_7	E	C_7	C_7^2	C_7^3	C_7^4	C_7^5	C_7^6		$\varepsilon = \exp(2\pi i/7)$
A	1	1	1	1	1	1	1	z, R_z	$x^2 + y^2, z^2$
E_1	1	ε	ε^2	ε^3	ε^{3*}	ε^{2*}	ε^*	(x, y)	(xz,yz)
	1	ε^*	ε^{2*}	ε^{3*}	ε^3	ε^2	ε	(R_x, R_y)	
E_2	1	ε^2	ε^{3*}	ε^*	ε	ε^3	ε^{2*}		$(x^2 - y^2, xy)$
	1	ε^{2*}	ε^3	ε	ε^*	ε^{3*}	ε^2		
E_3	1	ε^3	ε^*	ε^2	ε^{2*}	ε	ε^{3*}		
	1	ε^{3*}	ε	ε^{2*}	ε^2	ε^*	ε^3		

C_8	E	C_8	C_4	C_2	C_4^3	C_8^3	C_8^5	C_8^7		$\varepsilon = \exp(2\pi i/8)$
A	1	1	1	1	1	1	1	1	z, R_z	$x^2 + y^2, z^2$
B	1	−1	1	1	1	−1	−1	−1		
E_1	1	ε	i	−1	$-i$	$-\varepsilon^*$	$-\varepsilon$	ε^*	(x, y)	(xz, yz)
	1	ε^*	$-i$	−1	i	$-\varepsilon$	$-\varepsilon^*$	ε	(R_x, R_y)	
E_2	1	i	−1	1	−1	$-i$	i	$-i$		$(x^2 - y^2, xy)$
	1	$-i$	−1	1	−1	i	$-i$	i		
E_3	1	$-\varepsilon$	i	−1	$-i$	ε^*	ε	$-\varepsilon^*$		
	1	$-\varepsilon^*$	$-i$	−1	i	ε	ε^*	$-\varepsilon$		

3. The D_n Groups

D_2	E	$C_2(z)$	$C_2(y)$	$C_2(x)$		
A	1	1	1	1		x^2, y^2, z^2
B_1	1	1	−1	−1	z, R_z	xy
B_2	1	−1	1	−1	y, R_y	xz
B_3	1	−1	−1	1	x, R_x	yz

D_3	E	$2C_3$	$3C_2$		
A_1	1	1	1		$x^2 + y^2, z^2$
A_2	1	1	−1	z, R_z	
E	2	−1	0	$(x, y)(R_x, R_y)$	$(x^2 - y^2, xy)(xz, yz)$

D_4	E	$2C_4$	$C_2 (= C_4^2)$	$2C'_2$	$2C''_2$		
A_1	1	1	1	1	1		$x^2 + y^2, z^2$
A_2	1	1	1	−1	−1	z, R_z	
B_1	1	−1	1	1	−1		$x^2 - y^2$
B_2	1	−1	1	−1	1		xy
E	2	0	−2	0	0	$(x, y)(R_x, R_y)$	(xz, yz)

D_5	E	$2C_5$	$2C_5^2$	$5C_2$		
A_1	1	1	1	1		$x^2 + y^2, z^2$
A_2	1	1	1	−1	z, R_z	
E_1	2	2 cos 72°	2 cos 144°	0	$(x, y)(R_x, R_y)$	(xz, yz)
E_2	2	2 cos 144°	2 cos 72°	0		$(x^2 - y^2, xy)$

D_6	E	$2C_6$	$2C_3$	C_2	$3C'_2$	$3C''_2$		
A_1	1	1	1	1	1	1		$x^2 + y^2, z^2$
A_2	1	1	1	1	−1	−1	z, R_z	
B_1	1	−1	1	−1	1	−1		
A_2	1	−1	1	−1	−1	1		
E_1	2	1	−1	−2	0	0	$(x, y)(R_x, R_y)$	(xz, yz)
E_2	2	−1	−1	2	0	0		$(x^2 - y^2, xy)$

4. The C_{nv} Groups

C_{2v}	E	C_2	$\sigma_v(xz)$	$\sigma'_v(yz)$		
A_1	1	1	1	1	z	x^2, y^2, z^2
A_2	1	1	−1	−1	R_z	xy
B_1	1	−1	1	−1	x, R_y	xz
B_2	1	−1	−1	1	y, R_x	yz

C_{3v}	E	$2C_3$	$3\sigma_v$		
A_1	1	1	1	z	$x^2 + y^2, z^2$
A_2	1	1	−1	R_z	
E	2	−1	0	$(x, y)(R_x, R_y)$	$(x^2 - y^2, xy)(xz, yz)$

C_{4v}	E	$2C_4$	C_2	$2\sigma_v$	$2\sigma_d$		
A_1	1	1	1	1	1	z	$x^2 + y^2, z^2$
A_2	1	1	1	−1	−1	R_z	
B_1	1	−1	1	1	−1		$x^2 - y^2$
B_2	1	−1	1	−1	1		xy
E	2	0	−2	0	0	$(x, y)(R_x, R_y)$	(xz, yz)

C_{5v}	E	$2C_5$	$2C_5^2$	$5\sigma_v$		
A_1	1	1	1	1	z	$x^2 + y^2, z^2$
A_2	1	1	1	–1	R_z	
E_1	2	2 cos 72°	2 cos 144°	0	$(x, y)(R_x, R_y)$	(xz, yz)
E_2	2	2 cos 144°	2 cos 72°	0		$(x^2 - y^2, xy)$

C_{6v}	E	$2C_6$	$2C_3$	C_2	$3\sigma_v$	$3\sigma_d$		
A_1	1	1	1	1	1	1	z	$x^2 + y^2, z^2$
A_2	1	1	1	1	–1	–1	R_z	
B_1	1	–1	1	–1	1	–1		
B_2	1	–1	1	–1	–1	1		
E_1	2	1	–1	–2	0	0	$(x, y)(R_x, R_y)$	(xz, yz)
E_2	2	–1	–1	2	0	0		$(x^2 - y^2, xy)$

5. The C_{nh} Groups

C_{2h}	E	C_2	i	σ_h		
A_g	1	1	1	1	R_z	x^2, y^2, z^2, xy
B_g	1	–1	1	–1	R_x, R_y	xz, yz
A_u	1	1	–1	–1	z	
B_u	1	–1	–1	1	x, y	

C_{3h}	E	C_3	C_3^2	σ_h	S_3	S_3^5		$\varepsilon = \exp(2\pi i/3)$
A'	1	1	1	1	1	1	R_z	$x^2 + y^2, z^2$
E'	{1	ε	ε*	1	ε	ε*}	(x,y)	$(x^2 - y^2, xy)$
	{1	ε*	ε	1	ε*	ε}		
A''	1	1	1	–1	–1	–1	z	
E''	{1	ε	ε*	–1	–ε	–ε*}	(R_x, R_y)	(xz, yz)
	{1	ε*	ε	–1	–ε*	–ε}		

C_{4h}	E	C_4	C_2	C_4^3	i	S_4^3	σ_h	S_4		
A_g	1	1	1	1	1	1	1	1	R_z	x^2+y^2, z^2
B_g	1	−1	1	−1	1	−1	1	−1		(x^2-y^2, xy)
E_g	1	i	−1	−i	1	i	−1	−i	(R_x, R_y)	(xz, yz)
	1	−i	−1	i	1	−i	−1	i		
A_u	1	1	1	1	−1	−1	−1	−1	z	
B_u	1	−1	1	−1	−1	1	−1	1		
E_u	1	i	−1	−i	−1	−i	1	i	(x, y)	
	1	−i	−1	i	−1	i	1	−i		

C_{5h}	E	C_5	C_5^2	C_5^3	C_5^4	σ_h	S_5	S_5^7	S_5^3	S_5^9		$\varepsilon = \exp(2\pi i/5)$
A'	1	1	1	1	1	1	1	1	1	1	R_z	x^2+y^2, z^2
E_1'	1	ε	ε^2	ε^{2*}	ε^*	1	ε	ε^2	ε^{2*}	ε	(x,y)	
	1	ε^*	ε^{2*}	ε^2	ε	1	ε^*	ε^{2*}	ε^2	ε		
E_2'	1	ε^2	ε^*	ε	ε^{2*}	1	ε^2	ε^*	ε	ε^{2*}		(x^2-y^2, xy)
	1	ε^{2*}	ε	ε^*	ε^2	1	ε^{2*}	ε	ε^*	ε^2		
A''	1	1	1	1	1	−1	−1	−1	−1	−1	z	
E_1''	1	ε	ε^2	ε^{2*}	ε^*	−1	$-\varepsilon$	$-\varepsilon^2$	$-\varepsilon^{2*}$	$-\varepsilon^*$	(R_x, R_y)	(xz, yz)
	1	ε^*	ε^{2*}	ε^2	ε	−1	$-\varepsilon^*$	$-\varepsilon^{2*}$	$-\varepsilon^2$	$-\varepsilon$		
E_2''	1	ε^2	ε^*	ε	ε^{2*}	−1	$-\varepsilon^2$	$-\varepsilon^*$	$-\varepsilon$	$-\varepsilon^{2*}$		
	1	ε^{2*}	ε	ε^*	ε^2	−1	$-\varepsilon^{2*}$	$-\varepsilon$	$-\varepsilon^*$	$-\varepsilon^2$		

C_{6h}	E	C_6	C_3	C_2	C_3^2	C_6^5	i	S_3^5	S_6^5	σ_h	S_6	S_3		$\varepsilon = \exp(2\pi i/6)$
A_g	1	1	1	1	1	1	1	1	1	1	1	1	R_z	x^2+y^2, z^2
B_g	1	−1	1	−1	1	−1	1	−1	1	−1	1	−1		
E_{1g}	1	ε	$-\varepsilon^*$	−1	$-\varepsilon$	ε^*	1	ε	$-\varepsilon^*$	−1	$-\varepsilon$	ε^*	(R_x, R_y)	(xz, yz)
	1	ε^*	$-\varepsilon$	−1	$-\varepsilon^*$	ε	1	ε^*	$-\varepsilon$	−1	$-\varepsilon^*$	ε		
E_{2g}	1	$-\varepsilon^*$	$-\varepsilon$	1	$-\varepsilon^*$	$-\varepsilon$	1	$-\varepsilon^*$	$-\varepsilon$	1	$-\varepsilon^*$	$-\varepsilon$		(x^2-y^2, xy)
	1	$-\varepsilon$	$-\varepsilon^*$	1	$-\varepsilon$	$-\varepsilon^*$	1	$-\varepsilon$	$-\varepsilon^*$	1	$-\varepsilon$	$-\varepsilon^*$		
A_u	1	1	1	1	1	1	−1	−1	−1	−1	−1	−1	z	
B_u	1	−1	1	−1	1	−1	−1	1	−1	1	−1	1		
E_{1u}	1	ε	$-\varepsilon^*$	−1	$-\varepsilon$	ε^*	−1	$-\varepsilon$	ε^*	1	ε	$-\varepsilon^*$	(x, y)	
	1	ε^*	$-\varepsilon$	−1	$-\varepsilon^*$	ε	−1	$-\varepsilon^*$	ε	1	ε^*	$-\varepsilon$		
E_{2u}	1	$-\varepsilon^*$	$-\varepsilon$	1	$-\varepsilon^*$	$-\varepsilon$	−1	ε^*	ε	−1	ε^*	ε		
	1	$-\varepsilon$	$-\varepsilon^*$	1	$-\varepsilon$	$-\varepsilon^*$	−1	ε	ε^*	−1	ε	ε^*		

6. The D_{nh} Groups

D_{2h}	E	$C_2(z)$	$C_2(y)$	$C_2(x)$	i	$\sigma_{(xy)}$	$\sigma_{(xz)}$	$\sigma_{(yz)}$		
A_g	1	1	1	1	1	1	1	1		x^2, y^2, z^2
B_{1g}	1	1	−1	−1	1	1	−1	−1	R_z	xy
B_{2g}	1	−1	1	−1	1	−1	1	−1	R_y	xz
B_{3g}	1	−1	−1	1	1	−1	−1	1	R_x	yz
A_u	1	1	1	1	−1	−1	−1	−1		
B_{1u}	1	1	−1	−1	−1	−1	1	1	z	
B_{2u}	1	−1	1	−1	−1	1	−1	1	y	
B_{3u}	1	−1	−1	1	−1	1	1	−1	x	

D_{3h}	E	$2C_3$	$3C_2$	σ_h	$2S_3$	$3\sigma_v$		
A_1'	1	1	1	1	1	1		$x^2 + y^2, z^2$
A_2'	1	1	−1	1	1	−1	R_z	
E'	2	−1	0	2	−1	0	(x, y)	$(x^2 - y^2, xy)$
A_1''	1	1	1	−1	−1	−1		
A_2''	1	1	−1	−1	−1	1	z	
E''	2	−1	0	−2	1	0	(R_x, R_y)	(xz, yz)

D_{4h}	E	$2C_4$	C_2	$2C_2'$	$2C_2''$	i	$2S_4$	σ_h	$2\sigma_v$	$2\sigma_{v6}$		
A_{1g}	1	1	1	1	1	1	1	1	1	1		$x^2 + y^2, z^2$
A_{2g}	1	1	1	−1	−1	1	1	1	−1	−1	R_z	
B_{1g}	1	−1	1	1	−1	1	−1	1	1	−1		$x^2 - y^2$
B_{2g}	1	−1	1	−1	1	1	−1	1	−1	1		xy
E_g	2	0	−2	0	0	2	0	−2	0	0	(R_x, R_y)	(xz, yz)
A_{1u}	1	1	1	1	1	−1	−1	−1	−1	−1		
A_{2u}	1	1	1	−1	−1	−1	−1	−1	1	1	z	
B_{1u}	1	−1	1	1	−1	−1	1	−1	−1	1		
B_{2u}	1	−1	1	−1	1	−1	1	−1	1	−1		
E_u	2	0	−2	0	0	−2	0	2	0	0	(x, y)	

D_{5h}	E	$2C_5$	$2C_5^2$	$5C_2$	σ_h	$2S_5$	$2S_5^3$	$5\sigma_v$		
A_1'	1	1	1	1	1	1	1	1		$x^2 + y^2, z^2$
A_2'	1	1	1	−1	1	1	1	−1	R_z	
E_1'	2	2 cos 72°	2 cos 144°	0	2	2 cos 72°	2 cos 144°	0	(x, y)	
E_2'	2	2 cos 144°	2 cos 72°	0	2	2 cos 144°	2 cos 72°	0		$(x^2 - y^2, xy)$
A_1''	1	1	1	1	−1	−1	−1	−1		
A_2''	1	1	1	−1	−1	−1	−1	1	z	
E_1''	2	2 cos 72°	2 cos 144°	0	−2	−2 cos 72°	−2 cos 144°	0	(R_x, R_y)	(xz, yz)
E_2''	2	2 cos 144°	2 cos 72°	0	−2	−2 cos 144°	−2 cos 72°	0		

D_{6h}	E	$2C_6$	$2C_3$	C_2	$3C_2'$	$3C_2''$	i	$2S_3$	$2S_6$	σ_h	$3\sigma_d$	$3\sigma_v$		
A_{1g}	1	1	1	1	1	1	1	1	1	1	1	1		$x^2 + y^2, z^2$
A_{2g}	1	1	1	1	–1	–1	1	1	1	1	–1	–1	R_z	
B_{1g}	1	–1	1	–1	1	–1	1	–1	1	–1	1	–1		
B_{2g}	1	–1	1	–1	–1	1	1	–1	1	–1	–1	1		
E_{1g}	2	1	–1	–2	0	0	2	1	–1	–2	0	0	(R_x, R_y)	(xz, yz)
E_{2g}	2	–1	–1	2	0	0	2	–1	–1	2	0	0		$(x^2 – y^2, z^2)$
A_{1u}	1	1	1	1	1	1	–1	–1	–1	–1	–1	–1		
A_{2u}	1	1	1	1	–1	–1	–1	–1	–1	–1	1	1	z	
B_{1u}	1	–1	1	–1	1	–1	–1	1	–1	1	–1	1		
B_{2u}	1	–1	1	–1	–1	1	–1	1	–1	1	1	–1		
E_{1u}	2	1	–1	–2	0	0	–2	–1	1	2	0	0	(x, y)	
E_{2u}	2	–1	–1	2	0	0	–2	1	1	–2	0	0		

D_{8h}	E	$2C_8$	$2C_8^3$	$2C_4$	C_2	$4C_2'$	$4C_2''$	i	$2S_8$	$2S_8^3$	$2S_4$	σ_h	$4\sigma_d$	$4\sigma_v$		
A_{1g}	1	1	1	1	1	1	1	1	1	1	1	1	1	1		$x^2 + y^2, z^2$
A_{2g}	1	1	1	1	1	–1	–1	1	1	1	1	1	–1	–1	Rz	
B_{1g}	1	–1	–1	1	1	1	–1	1	–1	–1	1	1	1	–1		
B_{2g}	1	–1	–1	1	1	–1	1	1	–1	–1	1	1	–1	1		
E_{1g}	2	$\sqrt{2}$	$-\sqrt{2}$	0	–2	0	0	2	$\sqrt{2}$	$-\sqrt{2}$	0	–2	0	0	(R_x, R_y)	(xz, yz)
E_{2g}	2	0	0	–2	2	0	0	2	0	0	–2	2	0	0		$(x^2 – y^2, xy)$
E_{3g}	2	$-\sqrt{2}$	$\sqrt{2}$	0	–2	0	0	2	$-\sqrt{2}$	$\sqrt{2}$	0	–2	0	0		
A_{1u}	1	1	1	1	1	1	1	–1	–1	–1	–1	–1	–1	–1		
A_{2u}	1	1	1	1	1	–1	–1	–1	–1	–1	–1	–1	1	1	z	
B_{1u}	1	–1	–1	1	1	1	–1	–1	1	1	–1	–1	–1	1		
B_{2u}	1	–1	–1	1	1	–1	1	–1	1	1	–1	–1	1	–1		
E_{1u}	2	$\sqrt{2}$	$-\sqrt{2}$	0	–2	0	0	–2	$-\sqrt{2}$	$\sqrt{2}$	0	2	0	0	(x, y)	
E_{2u}	2	0	0	–2	2	0	0	–2	0	0	2	–2	0	0		
E_{3u}	2	$-\sqrt{2}$	$\sqrt{2}$	0	–2	0	0	–2	$\sqrt{2}$	$-\sqrt{2}$	0	2	0	0		

7. The D_{nd} Groups

D_{2d}	E	$2S_4$	C_2	$2C_2'$	$2\sigma_d$		
A_1	1	1	1	1	1		$x^2 + y^2, z^2$
A_2	1	1	1	–1	–1	R_z	
B_1	1	–1	1	1	–1		$x^2 – y^2$
B_2	1	–1	1	–1	1	z	xy
E	2	0	–2	0	0	$(x, y); (R_x, R_y)$	(xz, yz)

D_{3d}	E	$2C_3$	$3C_2$	i	$2S_6$	$3\sigma_d$		
A_{1g}	1	1	1	1	1	1		$x^2 + y^2, z^2$
A_{2g}	1	1	–1	1	1	–1	R_z	
E_g	2	–1	0	2	–1	0	(R_x, R_y)	$(x^2 - y^2, xy), (xz, yz)$
A_{1u}	1	1	1	–1	–1	–1		
A_{2u}	1	1	–1	–1	–1	1	z	
E_u	2	–1	0	–2	1	0	(x, y)	

D_{4d}	E	$2S_8$	$2C_4$	$2S_8^3$	C_2	$4C_2'$	$4\sigma_d$		
A_1	1	1	1	1	1	1	1		$x^2 + y^2, z^2$
A_2	1	1	1	1	1	–1	–1	R_z	
B_1	1	–1	1	–1	1	1	–1		
B_2	1	–1	1	–1	1	–1	1	z	
E_1	2	$\sqrt{2}$	0	$-\sqrt{2}$	–2	0	0	(x, y)	
E_2	2	0	–2	0	2	0	0		$(x^2 - y^2, xy)$
E_3	2	$-\sqrt{2}$	0	$\sqrt{2}$	–2	0	0	(R_x, R_y)	(xz, yz)

D_{5d}	E	$2C_5$	$2C_5^2$	$5C_2$	i	$2S_{10}^3$	$2S_{10}$	$5\sigma_d$		
A_{1g}	1	1	1	1	1	1	1	1		$x^2 + y^2, z^2$
A_{2g}	1	1	1	–1	1	1	1	–1	R_z	
E_{1g}	2	2 cos 72°	2 cos 144°	0	2	2 cos 72°	2 cos 144°	0	(R_x, R_y)	(xz, yz)
E_{2g}	2	2 cos 144°	2 cos 72°	0	2	2 cos 144°	2 cos 72°	0		$(x^2 - y^2, xy)$
A_{1u}	1	1	1	1	–1	–1	–1	–1		
A_{2u}	1	1	1	–1	–1	–1	–1	1	z	
E_{1u}	2	2 cos 72°	2 cos 144°	0	–2	–2 cos 72°	–2 cos 144°	0	(x, y)	
E_{2u}	2	2 cos 144°	2 cos 72°	0	–2	–2 cos 144°	–2 cos 72°	0		

D_{6d}	E	$2S_{12}$	$2C_6$	$2S_4$	$2C_3$	$2S_{12}^5$	C_2	$6C_2'$	$6\sigma_d$		
A_1	1	1	1	1	1	1	1	1	1		$x^2 + y^2, z^2$
A_2	1	1	1	1	1	1	1	–1	–1	R_z	
B_1	1	–1	1	–1	1	–1	1	1	–1		
B_2	1	–1	1	–1	1	–1	1	–1	1	z	
E_1	2	$\sqrt{3}$	1	0	–1	$-\sqrt{3}$	–2	0	0	(x, y)	
E_2	2	1	–1	–2	–1	1	2	0	0		$(x^2 - y^2, xy)$
E_3	2	0	–2	0	2	0	–2	0	0		
E_4	2	–1	–1	2	–1	–1	2	0	0		
E_5	2	$-\sqrt{3}$	1	0	–1	$\sqrt{3}$	–2	0	0	(R_x, R_y)	(xz, yz)

8. The S_n Groups

S_4	E	S_4	C_2	S_4^3		
A	1	1	1	1	R_z	$x^2 + y^2, z^2$
B	1	–1	1	–1	z	$x^2 - y^2, xy$
E	1	i	–1	–i	(x, y); (R_x, R_y)	(xz, yz)
	1	–i	–1	i		

S_6	E	C_3	C_3^2	i	S_6^5	S_6		$\varepsilon = \exp(2\pi i/3)$
A_g	1	1	1	1	1	1	R_z	$x^2 + y^2, z^2$
E_g	1	ε	ε*	1	ε	ε*	(R_x, R_y)	$(x^2 - y^2, xy)$ (xz, yz)
	1	ε*	ε	1	ε*	ε		
A_u	1	1	1	–1	–1	–1	z	
E_u	1	ε	ε*	–1	–ε	–ε*	(x, y)	
	1	ε*	ε	–1	–ε*	–ε		

S_8	E	S_8	C_4	S_8^3	C_2	S_8^5	C_4^3	S_8^7		$\varepsilon = \exp(2\pi i/8)$
A	1	1	1	1	1	1	1	1	R_z	$x^2 + y^2, z^2$
B	1	–1	1	–1	1	–1	1	–1	z	
E_1	1	ε	i	–ε*	–1	–ε	–i	ε*	(x, y);	
	1	ε*	–i	–ε	–1	–ε*	i	ε	(R_x, R_y)	
E_2	1	i	–1	–i	1	i	–1	–i		$(x^2 - y^2, xy)$
	1	–i	–1	i	1	–i	–1	i		
E_3	1	–ε*	–i	ε	–1	ε*	i	ε		(xz, yz)
	1	–ε	i	ε*	–1	ε	–i	–ε*		

9. The Cubic Groups

T	E	$4C_3$	$4C_3^2$	$3C_2$		$\varepsilon = \exp(2\pi i/3)$
A	1	1	1	1		$x^2 + y^2 + z^2$
E	1	ε	ε*	1		$(2z^2 - x^2 - y^2,$
	1	ε*	ε	1		$x^2 - y^2)$
T	3	0	0	–1	(R_x, R_y, R_z); (x, y, z)	(xy, xz, yz)

T_h	E	$4C_3$	$4C_3^2$	$3C_2$	i	$4S_6$	$4S_6^5$	$3\sigma_h$		$\varepsilon = \exp(2\pi i/3)$
A_g	1	1	1	1	1	1	1	1		$x^2 + y^2 + z^2$
A_u	1	1	1	1	–1	–1	–1	–1		
E_g	1	ε	ε*	1	1	ε	ε*	1		$(2z^2 - x^2 - y^2,$
	1	ε*	ε	1	1	ε*	ε	1		$x^2 - y^2)$
E_u	1	ε	ε*	1	–1	–ε	–ε*	–1		
	1	ε*	ε	1	–1	–ε*	–ε	–1		
T_g	3	0	0	–1	1	0	0	–1	(R_x, R_y, R_z)	(xz, yz, xy)
T_u	3	0	0	–1	–1	0	0	1	(x, y, z)	

T_d	E	$8C_3$	$3C_2$	$6S_4$	$6\sigma_d$		
A_1	1	1	1	1	1		$x^2 + y^2 + z^2$
A_2	1	1	1	–1	–1		
E	2	–1	2	0	0		$(2z^2 - x^2 - y^2, x^2 - y^2)$
T_1	3	0	–1	1	–1	(R_x, R_y, R_z)	
T_2	3	0	–1	–1	1	(x, y, z)	(xy, xz, yz)

O	E	$6C_4$	$3C_2(=C_4^2)$	$8C_3$	$6C_2$		
A_1	1	1	1	1	1		$x^2 + y^2 + z^2$
A_2	1	–1	1	1	–1		
E	2	0	2	–1	0		$(2z^2 - x^2 - y^2, x^2 - y^2)$
T_1	3	1	–1	0	–1	(R_x, R_y, R_z) (x, y, z)	
T_2	3	–1	–1	0	1		(xy, xz, yz)

O_h	E	$8C_3$	$6C_2$	$6C_4$	$3C_2(=C_4^2)$	i	$6S_4$	$8S_6$	$3\sigma_h$	$6\sigma_d$		
A_{1g}	1	1	1	1	1	1	1	1	1	1		$x^2 + y^2 + z^2$
A_{2g}	1	1	–1	–1	1	1	–1	1	1	–1		
E_g	2	–1	0	0	2	2	0	–1	2	0		$(2z^2 - x^2 - y^2, x^2 - y^2)$
T_{1g}	3	0	–1	1	–1	3	1	0	–1	–1	(R_x, R_y, R_z)	
T_{2g}	3	0	1	–1	–1	3	–1	0	–1	1		(xz, yz, xy)
A_{1u}	1	1	1	1	1	–1	–1	–1	–1	–1		
A_{2u}	1	1	–1	–1	1	–1	1	–1	–1	1		
E_u	2	–1	0	0	2	–2	0	1	–2	0		
T_{1u}	3	0	–1	1	–1	–3	–1	0	1	1	(x, y, z)	
T_{2u}	3	0	1	–1	–1	–3	1	0	1	–1		

10. The Groups $C_{\infty v}$ and $D_{\infty h}{}^{\Phi}$ for Linear Molecules

$C_{\infty v}$	E	$2C_{\infty h}{}^{\Phi}$	…	$\infty\sigma_v$		
$A_1 \equiv \Sigma^+$	1	1	…	1	z	$x^2 + y^2, z^2$
$A_2 \equiv \Sigma^-$	1	1	…	−1	R_z	
$E_1 \equiv \Pi$	2	2 cos Φ	…	0	(x, y); (R_x, R_y)	(xz, yz)
$E_2 \equiv \Delta$	2	2 cos 2Φ	…	0		$(x^2 - y^2, xy)$
$E_3 \equiv \Phi$	2	2 cos 3Φ	…	0		
…	…	…	…	…		

$D_{\infty h}$	E	$2C_{\infty}{}^{\Phi}$	…	$\infty\sigma_v$	i	$2S_{\infty}{}^{\Phi}$	…	∞C_2		
$\Sigma_g{}^+$	1	1	…	1	1	1	…	1		$x^2 + y^2, z^2$
$\Sigma_g{}^-$	1	1	…	−1	1	1	…	−1	R_z	
Π_g	2	2 cos Φ	…	0	2	−2 cos Φ	…	0	(R_x, R_y)	(xz, yz)
Δ_g	2	2 cos 2Φ	…	0	2	2 cos 2Φ	…	0		$(x^2 - y^2, xy)$
…	…	…	…	…	…	…	…	…		
$\Sigma_u{}^+$	1	1	…	1	−1	−1	…	−1	z	
$\Sigma_u{}^-$	1	1	…	−1	−1	−1	…	1		
Π_u	2	2 cos Φ	…	0	−2	2 cos Φ	…	0	(x, y)	
Δ_u	2	2 cos 2Φ	…	0	−2	−2 cos 2Φ	…	0		
…	…	…	…	…	…	…	…	…		

11. The Icosahedral Groups

I_h	E	$12C_5$	$12C_5{}^2$	$20C_3$	$15C_2$	i	$12S_{10}$	$12S_{10}{}^3$	$20S_6$	15σ		
A_g	1	1	1	1	1	1	1	1	1	1		$x^2 + y^2 + z^2$
T_{1g}	3	½(1+ √5)	½(1− √5)	0	−1	3	½(1− √5)	½(1+ √5)	0	−1	(R_x, R_y, R_z)	
T_{2g}	3	½(1− √5)	½(1+ √5)	0	−1	3	½(1+ √5)	½(1− √5)	0	−1		
G_g	4	−1	−1	1	0	4	−1	−1	1	0		
H_g	5	0	0	−1	1	5	0	0	−1	1		$(2z^2 - x^2 - y^2, x^2 - y^2, xy, yz, zx)$
A_u	1	1	1	1	1	−1	−1	−1	−1	−1		
T_{1u}	3	½(1+ √5)	½(1− √5)	0	−1	−3	−½(1− √5)	−½(1+ √5)	0	1	(x, y, z)	
T_{2u}	3	½(1− √5)	½(1+ √5)	0	−1	−3	−½(1+ √5)	−½(1− √5)	0	1		
G_u	4	−1	−1	1	0	−4	1	1	−1	0		
H_u	5	0	0	−1	1	−5	0	0	1	−1		

For the pure rotation group I, the outlined section in the upper left is the character table; the *g* subscripts should, of course, be dropped and *(x, y, z)* assigned to the T_1 representation.

Exercises

Questions

1. Define symmetry and symmetry elements.
2. List the distinct operations implied by C_6 axis.
3. Define symmetry operation. Mention any two symmetry operations.
4. Define centre of symmetry. Which of the molecules have a centre of symmetry?
   (i) C_6H_6 (ii) CO_2 (iii) SO_2 (iv) H_2O
5. List the symmetry elements present in the following molecules.
   (i) NH_3 (ii) CO_2 (iii) BCl_3 (iv) Benzene (v) H_2O (vi) B_2H_6 (vii) N_2F_2 (viii) PF_5 (ix) Naphthalene (x) CCl_3Br
6. Define the terms (i) class (ii) subgroup.
7. What are commutative operations?
8. Define the properties of a group.
9. Show that a S_n axis generates $2n$ operations if n is odd and n operations if n is even.
10. What is meant by an Abelian group?
11. Evaluate the products $\sigma_v \sigma_{v'}$ and $C_2 \sigma_v$ for a C_{2v} point group.
12. Prove that the S_n axis is the same as the $C_{n/2}$ axis when n is even.
13. Prove that the S_n axis is the same as the C_n axis when n is odd.
14. What are conjugate elements?
15. Define order of a group.
16. What is the necessity to include E with all the groups?
17. In C_3 and C_{3v} point groups, predict whether C_3 and C_3^2 are conjugate elements.
18. What is an isomorphic group?
19. The presence of $C_{2(x)}$ and $C_{2(y)}$ in a molecule automatically requires the presence of $C_{2(z)}$ in the molecule. Why?
20. Show that the following commute.(i) C_n and σ (ii) σ_{xy} and σ_{yz} (iii) $C_{2(z)}$ and $C_{2(y)}$
21. Identify the subgroups of the D_{2h} point group.

22. List the operations of $S_n^{\ n}$ when n is odd and when n is even.
23. $C_n^{\ n} = E$ but $S_n^{\ n} \neq E$ when n is odd. Why?
24. Show that S_n requires the presence of C_n when n is odd.
25. What is meant by the term group? Assign appropriate point groups to the following molecules.
    (i)BCl_3 (ii) Benzene (iii) C_2H_4 (iv)CH_4 (v) NH_3 (vi) H_2O_2 (*cis* and *trans*)
    (vii) $CHCl_3$ (viii) SO_2 (ix) H_2S (x) CH_2Cl_2 (xi) HCHO (xii) CH_2=C(Cl) (F)
    (xiii) Pyridine (xiv) Cyclohexane (xv) HOCl (xvi) HOD
    (xvii) ClF_3 (T-shaped) (xviii) XeF_4 (xix) $XeOF_4$ (xx) BrF_5 (xxi) Allene
    (xxii) C_2H_2 (xxiii) Ferrocene (staggered and eclipsed) (xxiv) Naphthalene
    (xxv) Anthracene (xxvi) Cyclopentadiene (xxvii) *Cis*-$Pt(NH_3)_2Cl_2$
    (xxviii) $[Co(en)_3]^{3+}$ (xxix)HCO_3^-(xxx)S_8 (xxxi)$SiF_6^{\ 2-}$ (xxxii)$CO_3^{\ 2-}$
    (xxxiii) *p*-Dichlorobenzene (xxxiv) NaOCl (xxxv) H_2
26. State the conditions that must be obeyed by a set of elements to form a group. Show that the set of elements 1, –1, *i* and –*i* forms a group.
27. What is meant by a similarity transformation? Carry out all possible similarity transformations on the elements of the C_{3v} point group.
28. Identify the classes and subgroups of the C_{3v} point group.
29. What is meant by matrix representation of a group? Arrive at the matrix representation for the C_{2v} point group.
30. Give the transformation matrix of (i) E (ii) σ_{xy} (iii) $C_{n(z)}$ (iv)S_n and (v) i
31. Show that $S_z = i$ using the matrix method.
32. Construct the group multiplication table for the C_{3v} point group.
33. What are reducible and irreducible representations?
34. Explain the block factoring method.
35. Carbon monoxide belongs to the $C_{\infty v}$ point group. Find the irreducible representation that results when symmetry operations of this group are performed on the z coordinate of the oxygen atom.
36. Reduce the given reducible representation in to its component irreducible representations of the C_{2v} group.

    $\Gamma = 4 \ \ 2 \ \ 0 \ \ 2$
37. The character of a reducible representation of a C_{2v} group are

    $\Gamma = 27 \ \ -1 \ \ 15$

    Determine how many times each irreducible representation of C_{2v} is present in Γ.
38. Define character. Construct the C_{3v} character table.
39. What is direct product? Give its application.
40. Differentiate direct product and matrix multiplication.
41. State the great orthogonality theorm and give its significance.
42. Obtain the reducible representation for the π-orbitals in 1,3,-butadiene.

43. How many irreducible representations are possible for the C_{3v} and C_{2v} point groups?
44. Find the representation corresponding to the total number of degrees of freedom of the water molecule.
45. Find the direct products of the following for the C_{2v} point group. $A_1 \times A_1$, $A_1 \times A_2$, $B_1 \times B_1$, $B_1 \times A_1$, $B_2 \times A_2$, $B_2 \times B_2$, $B_1 \times B_1$, $A_2 \times A_2$
46. Find the symmetry of stretching and bending vibrations of the water molecule.
47. How is group theory used to determine the IR activity of the fundamental vibrations of a molecule?
48. Discuss the hybridisation schemes for σ-orbitals using the tetrahedral molecule AB_4 and the D_{3h} AB_5 molecule.
49. Using group theory, how can one predict the infrared and Raman activity or otherwise of a normal mode of vibration of the NH_3 molecule.
50. Show how group theory helps in the prediction of IR activity of the fundamentals of HCHO.
51. Determine the symmetry of the $\vec{z}$ and R_z in a C_{3v} symmetrical environment.
52. Find the symmetry of $\vec{x}$ and $\vec{y}$ belonging to a C_{2v} point group.

MULTIPLE CHOICE QUESTIONS

1. Which of the following molecules possesses a centre of symmetry?
   a. $PtCl_4$ b. NH_3 c. Chlorobenzene (d) pyridine
2. The number of symmetry operations associated with the S_4 axis is
   a. 3 b. 6 **c. 4** d. 8
3. The point group for the benzene molecule is
   a. D_{2h} **b. D_{6h}** c. C_{3v} d. C_{2h}
4. Pick out the pair of conjugate elements for the C_{3v} point group
   a. C_3' and C_3^2 b. C_3' and σ_{v1} c. C_3^2 and σ_{v2}
   d. C_3^1 and σ_{v2}
5. The number of symmetry elements for a tetrahedral molecule is
   a. 15 **b. 24** c. 13 d. 20
6. The point group for the ethylene molecule is
   a. D_{2h} b. C_{2h} c. C_{3h} d. C_{4h}
7. The symmetry of the symmetric stretching vibration of the water molecule is
   a. A_1 b. A_2 c. B_2 d. B_1
8. The symmetry of the anti-symmetric stretching vibration of the water molecule is
   a. A_1 b. A_2 **c. B_2** d. B_1

Index